STRAWBALE HOMEBUILDING

EARTH GARDEN BOOKS
Trentham, Victoria

ACKNOWLEDGEMENTS

Strawbale Homebuilding is the result of many generous contributions from the following people:

Ahtee Chia
Andy & Daniela Donaldson
Anne Stelling & John Walker
Brian Cleeves
Danny Vonk
David & Heather Allworth
Des Menz
Dianne & Denis Heaney
Frank Thomas
Graham & Kaye Potter
Heather Marr
Janet Dunn & Stephen Butler
Jill Disley
Jo & Kerry Armstrong
John Glassford & Susan Wingate-Pearse
Josephine James
Julian & Carolan Castagna
Julie Gilfillan
Justin & Meg Byrnes
Katherine Perrin
Lachlan Burke
Lottie
Marcus Ward
Mareika & Harry Borchard
Margot Stephens & Paul Phillis
Matthew & Fiona Bunton
Michael & Shaina Hennessy
Michael and Ilana Cowan
Michael Kingsbury
Milosh Obradovich
Neil & Jacqui Cuthbert
Nicolas Goodwolf
Nik Vollmer
Per & Helen Bernard
Peter & Laurel Edwards
Peter Scott
Phillip Richards
Rose White & John Easterbrook
Sam Statham
Scott McGilchrist
Shane Naughton
Sharron Baker & Rudy Stoffel
Steffan Klein
Steve and Leontine
Steve Sainsbury
Sue & Phil Ogden
Sue Ewart & Don O'Connor
Vipassana Meditation Centre
Wayne and Fiona Gipters

EARTH GARDEN BOOKS
RMB 427 Trentham,
Victoria, 3458,
AUSTRALIA.
Phone (03) 5424 1819
Fax (03) 5424 1743.
Email: <earthmag@kyneton.net.au>.
World Wide Web site: <www.earthgarden.com.au>

A.C.N. 086 043 567
A.B.N. 69 086 043 567

ISBN 0 9586397-4-4

This book is printed on plantation-grown paper in the interests of preserving our planet's precious resources.

Edited by Alan T Gray and Anne Hall.
Graphic design by Tony Fuery.

Printed by Australian Print Group, Maryborough, Victoria.

Distributed by Gordon and Gotch, Burwood, Melbourne, and Gemcraft Books, Burwood, Melbourne.
Distributed in the USA by Chelsea Green, White River Junction, Vermont (800) 639-4099 <www.chelsea.green.com>.

FRONT COVER (clockwise from top): Neil and Jacqui Cuthbert's strawbale home near Orange in New South Wales is a model of energy efficiency and understated artistry. The Walker/Stelling home near Beechworth in north-east Victoria is the ultimate in stylish Australian Santa Fe elegance. The internal walls of the Vipassana Meditation Centre in the NSW Blue Mountains glow with a powerful serenity.

BACK COVER (from top to bottom): Wayne and Fiona Gipters' mountain-style home at Leura in the Blue Mountains. Lottie's $20,000 home in a lush Queensland bush setting. John Easterbrook and Rose White in front of their earthen spray-rendered home in Victoria's Otway Ranges.

FOREWORD

BUILDING A HOUSE out of strawbales is no longer the strange alternative idea that it appeared to be several years ago. Many people now know that this method has been used with great success in the USA for over one hundred years; that strawbale homes can have walls that are six or more times better-insulated than other wall materials; and that strawbale homes can be built quickly and cheaply — if you do the work yourself and become a dedicated scrounger and recycler.

Many people tell in this book how they were inspired to build in strawbale by the beautiful American book *The Straw Bale House*. We hope that *Strawbale Homebuilding* will be an adjunct to *The Straw Bale House*; that it will be a useful and encouraging guide to many people who want to build their own shelter, don't want a huge mortgage roped around their necks for 25 years, and who like the environmental and aesthetic benefits of strawbale building.

In May 2000, my wife, Judith, and our sons travelled up and down the East Coast of Australia photographing strawbale homes and interviewing their owners. Judith and I were staggered and inspired by the superb design innovations, the hard work and stunning results of people from a whole host of backgrounds and attitudes.

We liked the wisdom of a German-Australian builder we met near Sydney. Frank Thomas built Janet Dunn's superb pavilion-style home near Galston, and he comes from the background of German builders who must be expert and precise with all they build. Frank simply says that "any good builder should be able to build a strawbale house. If they can't, they are not a good builder". This, combined with the famous maxim from earth-building guru, Bob Rich: "If you can cook, you can build", should be enough encouragement for people to roll up their sleeves and start . . . with Helen Bernard's wonderful plans for a backyard strawbale chook shed. These plans start on page 143.

We are very grateful to the contributors to *Strawbale Homebuilding*. They have generously shared their information, tips, secrets, mistakes and costings. Five years ago there were one or two strawbale homes in Australia. Today there are around a hundred, and they're legal in every State.

Strawbale owner building is a contagious disease — watch out!

Alan T Gray

Alan T Gray
Trentham, Victoria.

Strawbale in Australia

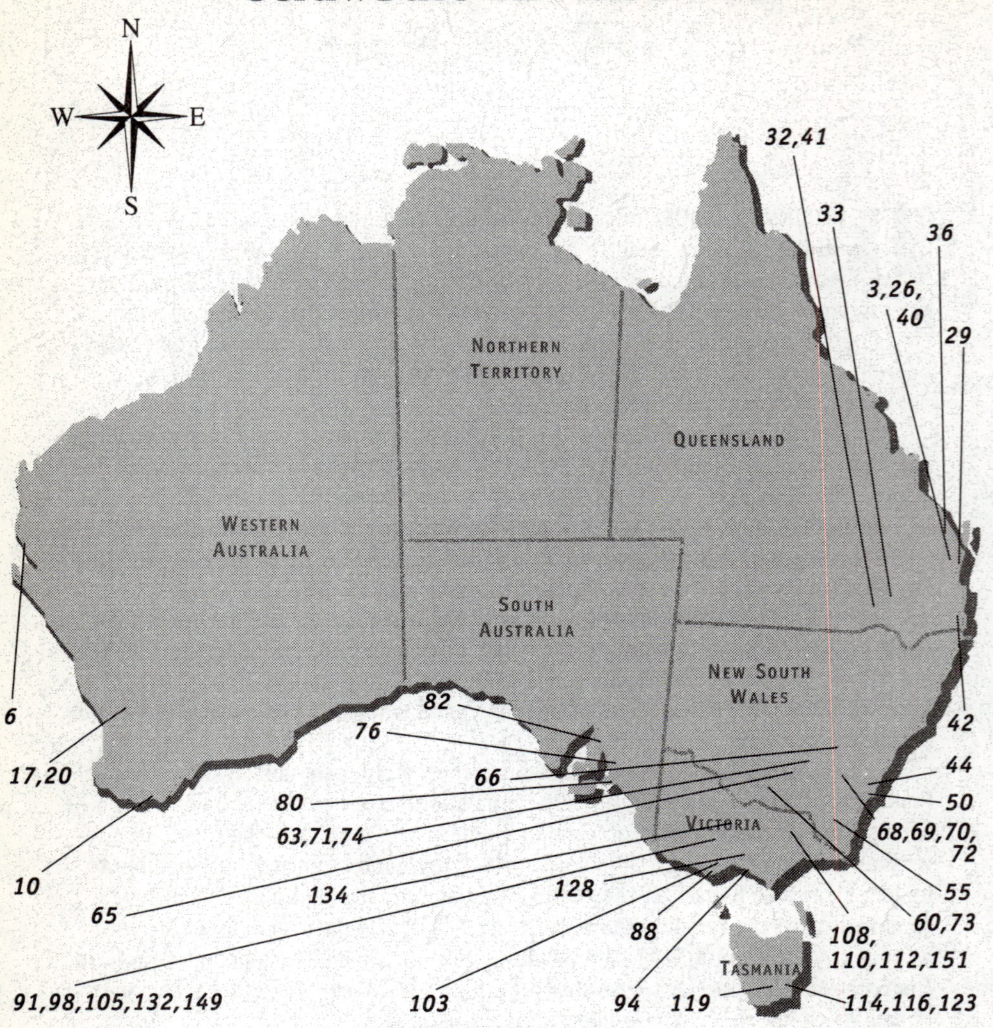

These page numbers refer to houses featured in the book.

CONTENTS

INTRODUCTION
Earth friendly building 2

WESTERN AUSTRALIA
Strawbale in the cyclone zone 6
Building the dream house 10
Round house in the WA bush 17
Strawbale thrives in the wheat belt 20

QUEENSLAND
Architectural philosophy leads to a beautiful home 26
From caravan to palace 29
Huge executive home 32
DIY metal frame resists termites 33
Strawbale in a hot climate 36
Suburban bliss with rock-solid walls 41
Determined to recycle with style 42

NEW SOUTH WALES
Attention to detail gives delightful result 44
The Bower built of straw 50
So natural that horses tried to eat it 55
Our top ten strawbale buildings 60
Modest appearance gradually reveals stunning character 63
Champagne views on a clever budget 65
Heavenly work space a strawbale showcase 66
Mountain-style home 68
Quiet elegance on Blue Mountains 69
Creative mountain home fires imagination 70
One day a machinery shed, next day a strawbale home 71
Perfect contemplation 72
Pioneers do house and land for $20,000 73
Pavilion-style home with a holiday atmosphere 74

SOUTH AUSTRALIA
First contract-built home in South Australia 76
Strawbale bears the load 80
House evolves from strawbale revolution 82

VICTORIA
Been there, baled that 88
Innovative materials need imaginative design 91
Four months and it's finished 94
We did it our way 98
Our strawbale philosophy 101
Winery builds country's biggest 103
Bed and breakfast in a cold climate 105
Australian-Santa Fe elegance 108
Professional owner-building 110
Grand biodynamic winery 112

TASMANIA
Strawbale house defies Tasmanian winter 114
With a little help from our (50) friends 116
And on the seventh year they rested 119
Constructive building 123

BUT HOW DO I . . . ?
Getting house plans approved 126
A council's perspective on strawbale houses 128
Life beyond brick veneer 131
Green house built of straw 132
Have render, will travel 134
Build your own chookhouse and greenhouse 143
Frequently asked questions 148
Contact list and further reading 154
Index 155

INTRODUCTION

Earth friendly building

Strawbale construction is far superior to many modern building techniques in terms of environmental impact.

by John Glassford and Susan Wingate-Pearse
Ganmain, New South Wales.

AS MUCH AS one-tenth of the total world economy is dedicated to the building and construction industry. In terms of materials used, this industry uses an even greater proportion. One-sixth to one-half of the world's wood, minerals, water and energy are used in the manufacturing and transport of construction materials.

In terms of energy use, there is an even more alarming statistic. Forty-five per cent of all the energy consumed in the world is used in the manufacturing and transportation of building and construction materials. This is more than all other uses combined, and is clearly unsustainable. (*World Watch Magazine*, Vol. 7, No. 6).

Strawbale building can help re-adjust this severe imbalance. For example, it costs the environment 6.15 million kilojoules of energy to manufacture one tonne of concrete. On the other hand it costs 119,250 kilojoules of energy to produce one tonne of straw. And one tonne of straw goes further than one tonne of concrete. (These calculations are by Richard Hoffmeister from the Frank Lloyd Wright School of Architecture in Scottsdale, Arizona, United States.)

HISTORY OF STRAWBALE BUILDING

As long as humans have been creating shelter, they have used straw and grasses in a variety of building methods to provide safe, dependable and comfortable housing in many climates and environments. These building methods include:

- straw-thatched bundles;
- straw mixed with earth or mud mortar, as in cob houses;
- straw walls which have been used in Spain and Belgium for hundreds of years;
- straw roofs and sod roofs.

In the United States, a new era of building with straw and grasses began in the late 1800s with the development of the steam baler. This made it possible to compress straw and hay into bales about 900 millimetres by 450 millimetres by 375 millimetres. It was in the State of Nebraska, a land that yielded magnificent stands of meadow hay, that people built the first strawbale buildings. Many of these houses still stand today.

Not many more strawbale buildings were constructed after the introduction of modern building methods, apart from a few houses and churches. In 1974 an article written by Roger Welsch of Nebraska was published in a book called *Shelter*. Welsch's article described a bale building. That article probably did more to initiate a revival of bale building than any other factor. Several strawbale houses were built as

Conondale engineer and designer, Michael Kingsbury, helped Ilana and Michael Cowan of the Crystal Waters Permaculture Village near Maleny in Queensland with their stunning strawbale and timber house.

a direct result of that article.

In 1987, David Bainbridge, co-author of *The Straw Bale House*, spoke at a permaculture design course in California on strawbale houses. Several people attending that course went on to build their own strawbale houses. The permaculture movement has been instrumental in promoting and encouraging people in many countries of the world to build with straw. So far more than 1000 houses have been built in the United States from strawbales. Other countries moving fast into strawbales are Canada, France, New Zealand, England, Scotland, Wales, Mexico, Mongolia and Australia.

Beauty of strawbale

Buildings made from strawbale have a particular beauty and comfort. The thickness of the walls and their subtle curves have a special character. The walls create an overall feeling of comfort, compared with thin walls. They are similar in appearance to old, thick, stone and adobe walls. The mass of strawbales induces physical and psychological feelings of goodwill and well-being.

Strawbale walls are a pleasure to look at and to touch, and give a soft, sound quality to each room. Naturally plastered, strawbale walls can breathe, resulting in indoor air that feels fresh, clean and invigorating. Within these silent, sculpted walls which impart a sense of timeless peace, a new and better vision of shelter resides.

EASE OF CONSTRUCTION

Building with strawbale requires less skill than conventional building materials. It is forgiving, and encourages individual creativity. It relaxes the whole construction process, and allows inexperienced and unskilled people the opportunity to become involved in creating their own houses.

The basic methods of strawbale construction can be learned in a three-day workshop. Everyone in the family can become involved in the construction. There are two main methods of strawbale construction:

Loadbearing: Bales are stacked in a running bond like bricks and pinned together. The walls directly bear and support the weight of the roof.

Bale infill: Another structural system supports and attaches the roof, and the bales are inserted as infill material.

ENERGY EFFICIENCY

The added mass of the bale effectively increases the thermal performance of the wall system. The increased insulation allows smaller heating or cooling systems to be installed. It also dramatically increases the efficiency of a passive solar construction, leading to substantial savings in gas, electricity and wood fuel bills. Strawbale walls are 30 times less energy intensive than wood frame walls to manufacture.

ENVIRONMENTAL BENEFITS

Farmers burn straw. This produces carbon dioxide and promotes deterioration of the ozone layer. Straw burning causes air pollution through smoke and particulate emission. Every 1.02 million tonnes of straw burned releases 57,102 tonnes of carbon dioxide. In the United States, 204 million tonnes of waste straw are produced every year. Annual straw burning in California produces more carbon dioxide and particulates than all the electric power generating plants combined in California. This air pollution has prompted California's authorities to ban straw burning.

If we can bale straw and use it to build, instead of burning it, we can create a win-win situation for cereal and cane growers, builders, house owners, and of course, the environment. New South Wales produces tremendous quantities of straw — more than one million tonnes from rice-growing alone. Let us assume that all of this rice straw could be used for building houses. Say the average building of 200 square metres uses 1000 small bales, that is 25 tonnes per house. If we used all that available straw, 40,000 single-storey houses could be built each year. Clearly the potential is enormous.

Therefore using strawbales for building could help control global warming and atmospheric deterioration. Removal of straw from the rice fields would substantially reduce methane emission from microbial decomposition. There would be a significant reduction in the demand for native timber in houses if they were built from strawbales instead of the timber which is so prevalent today. Building with strawbales also reduces the need for paints and solvents that adversely affect human health and the atmosphere. Rendered walls can be precoloured at rendering.

SUSTAINABILITY

Straw can be grown in less than a year in a completely sustainable production system. This compares with a timber plantation which can take years to grow enough timber to produce a frame for a house. Using straw as a sustainable, renewable material for building could be especially beneficial in countries where timber is scarce and straw is plentiful. Australia is such a country.

Strawbale building has many other benefits. Strawbale walls resist termites and are sound-proof. When rendered, they provide a fire-resistant house that if built correctly, could survive an Australian bushfire. For the first time we have a real choice to do something to reduce energy consumption in the building of our homes, places of work, play and education. Strawbale buildings are a step in the right direction in conserving energy in the building and construction industry.

WESTERN AUSTRALIA

Strawbale in the cyclone zone

Sue, her husband, Phil, and three children live on a remote part of the Western Australian coast. Although inexperienced and removed from services, they built a three-bedroom strawbale house.

by Sue Ogden
Carnarvon region, Western Australia.

WOULD WE BUILD with bales again? The answer is definitely yes. When you look at the whole process of building your own home the bale walls seem like the easiest part.

My husband, my three children and I live in a hot, dry, windy climate in a cyclonic region. We're two-and-a-half hours from the nearest town, up mostly a dirt track. Temperatures can reach around 48°C in summer and 178 millimetres is our average annual rainfall — but then you can get over 120 millimetres in one downpour off a cyclone. It's also a place where you can get winds of over 200 kmh. So, building with bales was a good option for quite a few reasons.

Getting the plans passed at first seemed to be a bit of a hassle as we were the first bale construction in the area and we have to meet the Building Code of Australia to withstand cyclonic conditions. The building surveyor insisted we have a structural engineer to design and approve our plans. We were fortunate that our structural engineer, Syd Weston, had been to a Bale Up workshop run by Rudy Stoffel and Sharron Baker so he was familiar with bale building and was keen to design and put his stamp of approval on it.

Footings

Syd Weston had a good idea for footings. To cut down on the amount of concrete used he designed a T-shaped footing. If you were living in a wetter climate you would need to have more concrete above the ground than we have. He's also designed the footings so we would have ten millimetres either side of the footings after we had rendered the bales so that we can keep an eye out for termite tracks.

The house is a post and beam, steel frame construction with bale infill walls. We put old sump oil and tainted diesel under the footings plus granite dust around the edges of the footings to discourage termites and so far we have had no sign of activity and let's hope that it stays that way.

We had many blunders along the way, like putting in the parallel truss after the roof trusses were on, but nothing that a sledge hammer couldn't fix (ha, ha) — the advantages of a steel frame.

The walls

We used a Bale Up construction technique of laying trench mesh on top of the bales around the loadbearing posts with a concrete mortar mix on top of them as it is much easier than trying to impale bales onto long lengths of reobar. We had reobar out of the footings to hold the first course of bales but then every second

The three children liked playing in the straw house.

course we did the above-mentioned technique. This creates a bond beam every second course of bales so it strengthens the walls and helps with the high winds.

Modifying the bales wasn't hard at all with the use of a bale needle made from steel rod with an eye and a point one end and a handle on the other. My husband just used a hand-saw which was fine for notching out bales to fit around loadbearing posts. We used a whipper snipper for fine-tuning bale walls when they were completed and for bull-nosing windows.

For all the gaps underneath windows and at the top of the wall, we just grabbed handfuls of straw and pushed as much straw as we could into place which saves on cement render and uses up leftover straw. Next comes the chicken-wire which isn't hard but I remember spending three or four days just making up fencing wire 'tent pegs' and hammering these into the bales to keep the chicken-wire tight.

We also spent a few days putting wire ties all the way through the bales and then twisting it together on the other side. It's necessary little jobs like these that make it all go together more securely. We had made our own soil cement bricks for internal walls so before we rendered we attached "Brick Tor" to the inside chicken wire where these walls would meet the strawbale walls as this helps tie it all together.

Rendering

Rendering would have to be the worst part of it all. Reid (a good friend who was a godsend with the rendering), Phil and I suffered from bad burns as lime becomes caustic when it is wet, so cover up well when dealing with it (boots, gloves, long shirts).

It took about two weeks non-stop to render the inside with three coats and the outside with two coats. We used a cement lime render. We had a bit of a problem with cracking due to our hot, windy, dry climate. Even when we did the colour coat on the outside in more favourable conditions we still have slight cracking, nothing too bad, and it quite adds to that rustic look.

We used a ratio of nine parts sand to three parts lime to one part cement. We varied it slightly sometimes but we would have liked some advice. Trying to get advice from the plasterers in our nearest town or even Perth seemed a bit pointless as none of the people I spoke to had ever rendered strawbales. If we were to build with bales again I would try to get someone who had a cement spraying machine — it would be worth it. Otherwise beg, borrow or steal whatever labour you can and get a whole heap of people to help render as quickly as possible and get it out of the way.

REFLECTIONS

The advantages of building with bales are many but it all takes time, no matter what medium you choose to build with. It took us about seven months to level the site, do the footings (internal footings for a three-bedroom house), put up the steel frame, put the roof on, make the soil cement bricks, put the walls up, put the windows in and do the rendering.

We had a lot of help with the footings for which we'll be forever grateful. Due to our isolation the process took a lot longer than if you were somewhere where you didn't have to cart everything, including water and sand, and we couldn't get a cement truck in to pour the footings. One advantage of strawbale is that you can move in before the walls are rendered and you can be quite protected from the elements as your walls are up.

For plumbing requirements we pushed sleeves of plastic conduit (poly pipe rubber hose, whatever you have lying around that's big enough to accommodate copper pipe) through the walls where we thought we would need it so then the plumber could come at a later date and do all the plumbing. This way the plumber can weld without fear of fire as your walls are already rendered and any left over holes in walls can soon be plugged up with render or silicone.

As for electrical requirements: because we have soil cement brick internal walls and we run on solar power we decided we would just run wires down through the roof then notch the conduit into the internal walls, to save the time and effort of trying to attach power points to bale walls.

INSULATION

We have found the house extremely cool during the day on warm days and lovely and warm at night in winter, so that we don't need winter pyjamas anymore. We are looking forward to summer to see how cool it is. While we were building over summer there was a good six degree difference in temperature from our makeshift camp to the strawbale house and that was without the eaves being sealed.

As for it handling cyclonic winds, well I hope that we never get to experience the winds that Exmouth did (290 kmh) but if we ever do get a cyclone through here, I know where my husband and I would prefer to be and it's not in town sheltering in a tin shed, it will be among our bale and brick walls and we will leave the rest in the hands of God.

So for anyone contemplating the strawbale experience — go for it. The walls have a lovely feel to them and they look especially great at night under candle light, so much so that you don't want to hang anything on them to destroy that look or feel. As my husband often says, if inexperienced people like us can build a house then anyone can. And thanks once again to all those friends and acquaintances who helped us.

The Bale Up company run by Sharron Baker and Rudy Stoffel in Toodyay, Western Australia, have a great booklet which is reasonably priced and offers some good points as well as detailing the technique of putting trench mesh on top of bales instead of reobar through them. It's worth the $20 or so that it costs.

Sue says wrapping the walls with chicken-wire wasn't hard but she does recall there were days of preparation before the walls were ready for rendering.

The rendered walls. Rendering was by far the worst part, says Sue, and made harder by a lack of advice.

Building the dream house

A showcase home that attracts widespread attention was built with the help of family, friends, a workshop and tradespeople.

by Heather Marr
Denmark, Western Australia.

AFTER LIVING for six years in a tiny shed behind the small restaurant we run on our farm on the south coast of Western Australia, my husband, Malcolm, and I were really feeling our cramped conditions. We explored a number of possibilities for easing the problem with anyone who would listen, we got the tape measure out frequently and we compiled countless wish lists.

After one such session with tape measure, pencil and paper, my sister who was on holiday from Perth, announced: "You could just about build a whole new house for what you're planning and I know where you should build it!" Thus began our strawbale project. As with all such projects begun with high expectations and in almost complete ignorance of the scale of the task we had set ourselves, there have been high points and some distinctly low points.

We have gained some new skills, learned how generous people can be and have been taught some hard lessons in skepticism when dealing with tradesmen and suppliers and the 'warm fuzzy' myths that have grown up around strawbale building.

Working with my designer sister, Diana, and builder brother-in-law, Dick, has opened our eyes to the care and sheer attention to detail required to obtain the sorts of finish illustrated so temptingly in the colour plates of *The Straw Bale House*.

Why did we choose to build in strawbale when there are many fine examples of other building materials to be found in this area? Well, firstly there was the look of the buildings constructed using this method. The photographs of completed houses in *The Straw Bale House* book showed buildings which appealed to us because of their tranquil beauty, with deeply recessed windows and massive walls. We were particularly struck by the apparent simplicity of the construction methods which would enable us, two notoriously unhandy people, to be actively involved in the making of our house.

The environmental advantages of using a waste product and building a house which would require reduced energy inputs for comfortable living was also very appealing. There was also the aspect of pioneering a new building method which, if successful, could mean that others in our area would be able to build comfortable houses at a reasonable cost.

THE PLAN

Our house has really been the result of input from family and friends. The site, on a north-west facing slope with pasture and Karri forest behind was surveyed for us by a cousin. He also produced a site plan. Other family members donated weekends to help build and erect the internal stud framework and many hands have helped scrape paint from recycled windows, stuff lumpy walls with straw, make bales and provide encouragement.

Getting there: the walls are up, the roof is on, now for the rendering.

My sister Diana agreed to design our house for us. Having inspired the initial decision to build, she was keen to explore the possibilities of designing with bales, making use of their structural qualities as well as their thermal properties. The result is a spacious two bedroom house with a study and open living areas. Several internal walls are built with bales, although those in the bedrooms are standard stud walls.

There is a tower above the entry with a small loft behind, now known as 'Heather's Folly'. The tower walls which are approximately 5.5 metres high are constructed from bales. The tower adds interest to the roof-line and helps bring light into the house. We also believe that it will assist air circulation throughout the house. There are lots of windows, both to let in light and air and to take advantage of the forest and valley views. The house faces north to make the best use of passive solar conditions.

The house is a modified post and beam construction on a concrete raft slab. There were several reasons for the decision to use this method. Firstly, we needed local government approvals as speedily as possible as we were hosting a strawbale workshop in March 1998. The building inspector at the (then) Shire of Albany was interested enough in strawbale building to attend a workshop in Perth. On his advice, we opted for non-loadbearing walls as approval would be more straightforward.

According to *The Straw Bale House*, a post and beam building would not necessarily use more timber than one with loadbearing walls and there was less time needed for the walls to be compacted before the render coats were applied. A full set of plans and specifications was submitted to the Shire of Albany. These included a structural engineer's certificate confirming the strength of the building framework.

Approval was granted readily and site works commenced.

GETTING STARTED

Barley strawbales were ordered in December 1997 and began to be delivered soon afterwards. They were stored under cover. When we ordered them, we based our planning on bales of standard dimensions of 900 millimetres long by 450 millimetres wide by 300 millimetres high and tightly tied. Unfortunately, many bales did not meet these dimensions which meant a lot more work notching them to fit into the stud frames, or retying them to produce smaller bales. This process was assisted by having a sturdy, homemade bale press.

We elected to use Termimesh as our termite treatment in preference to chemical treatments. We will need to maintain the slab edge clear for a height of 75 millimetres and inspect it regularly but this seems a small price to pay to avoid traditional chemical treatment.

The first step in preparing the site came in January 1998 with the provision of power and telephone lines, a saga in itself. The sand pad was prepared without the need to bring in loads of compaction sand. As the site was levelled, several layers of terraces were constructed behind the house site. After compacting the pad, formwork was set up.

The concrete was poured and 750 millimetre lengths of ten millimetre diameter steel rod were embedded in the concrete at 450 millimetre intervals. These were withdrawn as the concrete cured to ensure the safety of people while they worked on the house. Once building started the rods were reinserted and topped with empty aluminium cans to shield their sharp edges.

The cans fluttered gaily with each breeze and generally made enough noise to keep our sheep off the building site before it was fenced. The steel rods were placed so as to spike each bale in the bottom course of bales twice. As the concrete cured it was damped down using tanker units borrowed from our local volunteer bushfire brigade (in return for a modest donation to brigade funds).

Parallel base boards were fixed on the edge of the slab and where internal bale walls were to go. These were infilled with blue metal to lift the bales off the concrete to keep them dry if we ever had washing machines or sinks overflowing. Prefabricated double stud frames built from treated plantation pine were then attached to the slab, through the base boards. The studs were braced temporarily to ensure that they remained square. It is these stud frames which support the roof, along with a large pole in the living area and one other metal pole built into one corner. The stud frames were placed at 1800 millimetre intervals. A continuous pitching beam was fixed to the internal face of the studs.

WORKSHOP

On the following weekend, a strawbale building workshop was held on the site. Unfortunately, this also coincided with the annual break of the season so it was a somewhat wet and windy affair. The weather cleared on Sunday and workshop participants were able to erect the bale walls on the second bedroom at the back and side of the house. Both walls contained windows so this would give people a good opportunity to see how these were finished.

While the bales were going up, friends slaked the lime for the render. This is a spectacular (and dangerous) process as quicklime explodes as it absorbs water and gives off great heat. The secret is to add small amounts of lime to 44 gallon drums half-filled with water, stirring frequently.

It is vital to wear safety glasses and to avoid leaning over the drums. More water can be added to achieve the desired consistency of the lime putty. Ideally, the lime-water mix should be allowed to sit for seven days or at least until it has cooled.

We used bags of hydrated lime rather than premixed lime putty as it is cheaper, and we could mix the necessary batches as required. Open drums of quicklime are potentially lethal to birds and small animals and dangerous to children and must be kept covered when not in use. When mix-

Heather and Malcolm wanted a high-quality finish for the interior and found a plasterer to assist them.

ing the render, we decided not to use any concrete as a lime-sand render will eventually set very hard indeed. But we used concrete on the walls in wet areas.

We discovered that there are drawbacks to having many enthusiastic helpers. It was very difficult to ensure adequate supervision of all the work and the walls went up unevenly. At the corner the walls developed a distinct outward lean. One window surround was covered with diamond mesh before the sill had been properly waterproofed. The small areas rendered were finished unevenly as it had been decided to apply the render directly to the bales, without first wrapping the walls with chicken-wire. The render did not stick very well, which was frustrating and, except for an area plastered by a plasterer who attended the workshop, the effect was very rough indeed.

Setbacks

With the break in the weather and autumn well on its way, work on the house slowed right down. As more walls went up they were covered with brightly-coloured tarpaulins. Unfortunately, we had greatly under-estimated the time required for building and the level of skill needed to undertake the project. The roof line was fairly complicated so we decided to get a tradesman to do the work.

However, the roofer we approached to erect the roof, having quoted for the job, kept putting us off. By the middle of winter it became clear that he was unwilling to do the work so we had to find another tradesman and go through the process of getting new quotes. Although the tarpaulins protected most of the walls and the bales seemed to shed much of the rain, two walls became wet through and needed to be pulled down when building recommenced in earnest in spring. A family of frogs was discovered in residence and they were not happy with their eviction.

The walls also faced threats from a visiting cow whose appetite was not thwarted by the fence around the building site, and not content to dine alone, she introduced our old horse to the pleasures of "off the wall dining". The two of them managed to create some serious dents in the regularity of the back walls.

Having sat forlornly through the winter with the frames in place and draped in tarps, it was a rather soggy house site that confronted us each day. Rather than let things deteriorate further, Dick decided to come down and work full time on the house, with our help when available.

THE ROOF

The first step was to complete the roof framework over the bedrooms. Never having constructed a roof frame before, Dick equipped himself with a little red book which gives the angles for cutting rafters. Fixing the first rafter took several hours but as he, Malcolm and a friend worked on, they gained confidence. It is remarkable how the roof framing made things look more defined. With the frame in place, the roofer was able to complete the first part of his work.

On the advice of a strawbale builder from Victoria we have made sure that we have put R 3.0 insulation in the ceiling over the living areas. Because of the super-insulating properties of the bales, summer heat entering the house through the roof can be difficult to vent. Good roof insulation and cross-ventilation through the house are important for summer comfort.

The flimsy looking framework began to look more like a house, with a recognisable shape and feel. The framework over the living areas was somewhat more complicated as the main beams of the roof hips were to be supported in the centre of the space by a tall steel pole. Another pole was required to support the verandah roof.

After discussing the options we decided to see if we could obtain a large log to form the roof support. We found just what we needed, a six-metre long piece of jarrah in a local wood yard. Having come from the demolished Deepwater Jetty in Albany it is not only a massive piece of saltwater seasoned wood, but it is truly part of the history of Western Australia. There are even borer holes to show its marine provenance! The same wood yard also supplied several smaller pieces of recycled wood for the verandah, and the roof was remodelled to make use of them.

Getting the large pole in place was the subject of much discussion and some hugely complicated schemes involving ropes, pulleys, lots of people and a high degree of risk! Fortunately, a neighbour solved the problem very simply by arriving one morning with his tractor and front bucket, some strong rope and scaffolding. Within five minutes, the log was in place in its metal base plate screwed to the floor and braced temporarily while cuts were made to accommodate the roof beams.

THE WALLS

Once the second section of roof was completed, work began again on the walls in the main living areas and the tower. This went on steadily with constant use of spirit levels to ensure that the bales were laid as evenly as possible and that the walls remained straight. Where necessary the bales were notched using a chainsaw. The dust and debris from the bales put a lot of stress on the chainsaw and it needed regular maintenance.

We tried trimming the walls with a whipper snipper but even with the blade attachment this was difficult to handle and produced very uneven results. We eventually invested in a petrol-driven hedge trimmer which was far more satisfactory and much easier to use.

As the walls went up they were pinned using metre-long bamboo rods. Two were inserted into each bale to attach it to the two courses beneath it. Each course was also pinned at corners with metal U-pins. Water-proofing was wrapped around the base of each window reveal and the edges of the reveal covered with diamond mesh.

Once the walls were completed we decided that — although expensive and time consuming — they should be wrapped with chicken-wire to ensure the best material for the render to adhere to. Our aim was for smooth and even walls with sharp,

The three coats of render to the outside walls are almost complete.

clear corners. The bales were trimmed carefully and any gaps stuffed with loose straw.

Plumbing pipes running through the bales were encased in plastic pipes for further protection from leaks. Where taps come off bale walls we have brought them up from the slab through cupboards, rather than fix them to the render. Fittings in the bathroom are fixed to the stud walls. Electrical wiring was run under the chicken-wire and the sites of switches and power points carefully noted. So far we have only 'lost' one point under the render and it took a serious search to find it again.

Rendering

The drums of lime putty left over from the workshop were still usable and because the remaining bags of hydrated lime had been stored carefully their contents were also usable. Because of the size of the house we investigated mechanical methods of applying the render.

We purchased an air texture gun which ran off a small compressor. Unfortunately, the gun kept clogging and was difficult to manoeuvre under the roof and at odd angles so we decided to go back to hand rendering. Although slow, our efforts improved rapidly and we were able to obtain quite a smooth finish. A young plasterer who had heard of our project joined us and rendering moved on apace.

The walls received three coats of render, both inside and out. The base coat is a thick one, to even out any unevenness in the walls. After a few days curing, the render was scratched to enable the next coat to stick. The render removed in this process is reused. The second and third coats are thinner, and the final one contains colour.

The bottom outsides of the window sills were rendered with a cement render and then carefully built up with lime render. Squaring the heads of windows and doors is a difficult task, certainly having the services of an experienced plasterer has made a big difference to the standard of the finish. Obtaining the straightest finish is a drawn out process but the results inside the house have been well worth the extra effort and expense. The final external coat will be somewhat

The house is designed to take advantage of the beautiful, surrounding landscape.

rougher than that on the internal walls.

A few tips

Well, what have we learned? I would summarise our strawbale experience in the following way:

- Start small. We would have been better off building a much smaller building such as a stable or a shed to gain the hands-on knowledge of the building techniques and to give us a guide to the real expenses involved.
- Work out just how smooth the final look should be and if necessary, set aside extra money for a professional plasterer to do the last coats. The difference in finish obtained by a good amateur and a fully trained tradesman is significant.
- Be prepared to supervise groups of willing helpers carefully or to work more slowly with only one or two helpers. Mistakes can be quite difficult and time consuming to correct and will also cause problems later on.
- Be realistic about the time you can give to building. The walls are relatively easy to build but they are only a small part of the final building. If we did this again we would set aside at least three months to devote solely to building. Because of our business and work commitments we have had to employ others and this has increased the costs substantially.
- Don't be afraid to use qualified tradesmen if necessary. Most of those who have worked with us, although initially sceptical, have become very supportive and have given good advice.
- Make friends with your credit union. Ours has been very supportive and we have been able to increase our mortgage to meet cost blowouts.
- Be prepared for lots of visitors to the site. When word gets out there is a lot of interest and people will come from all over the place to find out more.
- Look after your tradespeople and suppliers. As interest grows in strawbale building in your area, you will find you are asked to recommend suppliers and tradespeople. You may well find that tradespeople are willing to do you good deals in return for a little bit of promotion through your 'showcase' home.

Round house in the WA bush

Dianne and Denis wanted a beautiful and environmentally friendly building method for their new weekender.

by Dianne and Denis Heaney
Mullaloo, Western Australia.

FOR THE PAST year we have been putting together our strawbale house in the bush at Toodyay, which is about a 90-minute drive from Perth. Our property has many types of gum including jarrah, salmon and wandoo, families of kangaroos and other native wildlife. Many species of wildflowers are in abundance during spring. We have not fenced our property so wildlife can move freely between our land and Julimar Forest which adjoins it. The soil is gravelly with some rocky outcrops and a valley with a winter creek.

Our strawbale house is close to being finished and initially our plan is to use it as a weekender. For us the planning was exciting and we have learnt amazing things about building — and ourselves. We were coming from the city way of liv-

The round house, surrounded by hectares of bush, will first be used as a weekender.

Dianne and Denis found the beautiful leadlight windows and doors in salvage yards.

ing to seeing another way with endless possibilities for the future for our family and their families.

We loved the creativity which was available with this kind of building. We collected a number of doors and windows from salvage yards. The house was designed around these with the help of a Toodyay man named Rudy who also helped in the planning and building along with plumbing and other input.

The house is certainly unique, being round and split level with a turret to let in light at the top of high ceilings in the living area, which has a round kitchen in the centre surrounded by a very large half circle living and dining area. A feature of this area is the round fireplace in the middle of the living space with a copper flue overhead that hangs from a chain. We have wonderful wide window ledges and we will never need curtains as we have the bush all around us.

The house has two bedrooms upstairs, one has an ensuite with a spa. On the other side of the living area upstairs there is a second bedroom and another bathroom off the passage that will be used as a laundry. We have hand-painted tiles for both bathrooms to match the decor and the stained glass windows.

Our house is very well insulated. The bale building weekend was one of the greatest fun times we have shared with our friends.

We camped, worked and celebrated together, all 20 of us, under the guidance of Rudy, who was the strawbale expert. We had all learned so much by the end of the weekend, especially Rudy, who decided never to have so many people to teach at once again. It was such a hands-on experience for all of us. The amount that we achieved was amazing and we had help from friends who had electrical experience, along with friends who were farmers

This huge trunk was found lying on the forest floor of the Heaneys' property.

and others who had never done anything like this before. It will be something we will treasure for the rest of our lives.

This, for us, is something we would never have believed we could do and now we know that most things we can do either by ourselves, in consultation with others, or by finding an expert to do it for us — at a price. We have been the seed planters in our family and it will be up to others to continue and build on to what we have begun.

We intend slowing down and enjoying the tranquillity of the bush. You can have your dreams come true despite difficulties. In fact I believe having difficulties keeps you mindful of all that we have to be grateful for and that we are here to protect the Earth and all that lives and breathes from it.

Strawbale thrives in the wheat belt

Rudy and Sharron gained experience building their home before starting their business.

by Sharron Baker and Rudy Stoffel
Toodyay, Western Australia.

IN 1992 we finally made the decision to turn our backs on city life. In pursuit of our dream to be self-sufficient and mortgage-free we bought a bush block in Toodyay, 100 kilometres east of Perth. The block is quite rugged and on top of a north-facing hill with undulating open woodland. It commands wonderful views across the valley and farmlands. We knew we wanted to spend the rest of our days here.

We also knew that we needed to buy an earth-moving machine because of the steep, sloping contours. We felt that this would be more economic in the long run. It would also be a wonderful help when we developed our gardens. First we needed a driveway, as well as a house pad and access to the property. We put our order in and got a backhoe that we named 'Rosie'. Next we built a shed around the caravan that we had brought up from Perth. The shed was built out of rocks, rammed earth and the neighbours' off-cuts. Our first priority was to have a comfortable sleeping area with a good bed and a source of hot water. These were essential as we knew we would be working hard for a couple of years at least.

We had a combination of solar and slow combustion stove for heating and hot water and one 10,000 litre rainwater tank. We were then established enough to plan our new house. We agreed on a Swiss chalet mainly because Rudy is Swiss and he is familiar with this style of home. We designed an open-plan ground floor and mezzanine loft for the sleeping area. We decided to build into the hill to minimise the earthworks.

DISCOVERING STRAWBALE

Like most owner-builders we lack funds but we have millions of rocks on our property so we decided to use rocks to build the lower level of the house. This took about 18 months, mostly weekends and during the week: rocks in the morning and rocks after work, and there were those who thought we had rocks in the head, as well! We had nearly finished this stage when a friend visited and was excited about a new book he had bought. One glance through it and we became hooked. *The Straw Bale House* book came to be our bible.

We participated in a strawbale workshop where the project was to build a round chalet. The slab and base anchors were already in place and we learned how to push the bales over the reobar, how to make halfbales, and fit and sew the chicken-wire. After the lunch break we had a go rendering the walls of a cottage that was already built. It's fun to apply the render by hand and most had a ball

Rudy works on the strawbale walls for the upper storey.

doing it. Our preference is to render with a trowel because this gives a more professional finish. This was our introduction to the basics of strawbale construction and we came away inspired.

First we had to find a local farmer who was willing to supply us with strawbales to build our house. Back in 1995 this was a very novel idea and you can imagine the good natured ridicule we received, but nevertheless we came across one chap who thought we were just crazy enough to pull it off, and taking pity on us he gave us bales at $1 each.

Building

The upper section of the house had been designed for weatherboards. We had used structural pine studs and beams to accommodate these. The bales fitted nicely between the studs that had a spacing of 900 millimetres. To reduce the wall size to the dimension of the rock walls we laid the bales upright, with the strings facing outside, to achieve a 450 millimetre rendered wall instead of 550 millimetres.

It was a great advantage with Rudy being a plumber as the pipe work could be installed as we went along. We have found that the plastic compression pipes and fittings are the best for plumbing as there is no welding, which can be a tricky job around strawbales. The electrical cables were placed in conduit and these too were installed as we went. The wet areas were rendered, and coated with a rubber membrane sealant.

During the course of building we did many experiments with the bales in the form of laying them flat, sideways, lengthwise and with, and without, chicken-wire. We have made straw batts for the ceiling and thin vertical batts for internal walls. We have also experimented with various types of renders and patterns: adobe; bagging; limestone and sandstone. We also experimented with adding different soils for different colours to the mix. In this way we know what works best for us, and this has helped us immensely in helping others.

We have always been keen to try out new ideas, some have worked, some have not, but we have always learned something and through our experiments have built a unique home that has evolved from the ground up. From family, friends and strangers we gratefully received building materials with wonderful stories to tell of past lives as old farm sheds and turn of the century cottages. We were also given enough old corrugated sheeting to use as a temporary roof which protected the building materials from the elements. When we could afford it, we put a new roof over the top of it all, giving us a tropical roof with a 100 millimetre air flow. This is a bonus because during the summer we regularly get temperatures over 40°C.

BUILDING BUSINESS

During the past five years that we have been building, we have received a lot of attention from owner-builders and many others who have an interest in this method of building. We were able to show people exactly what a strawbale house looks and feels like, and what can be done with different renders. It was because of this growing interest that we started our business, Bale Up.

We began by running workshops from home which were very popular. We then progressed into designing and planning strawbale constructions and as the demand grew, moved into assisting owner-builders with construction work. Over the past two to three years we have been involved in a number of projects throughout Western Australia. These have varied from design and planning to assistance in the building of private homes, weekenders, studios and one stable. We are currently working between two projects: the first is a 14-metre diameter round house that was designed by Rudy. It is split-level and has two bedrooms, one with an ensuite with spa bath, the other with a separate bathroom, and a laundry all on the upper level. The lower level contains the kitchen, dining area and living areas.

The centre of the house rises to a turret which is supported by one six-metre high log with two smaller tree trunks as corner posts. All timbers have been gathered from the owners' property, from the forest floor where they had fallen. This house is rugged and rustic with beautiful French doors and wonderful old leadlight windows (see page 17).

SHED CONVERSION

The other project is the conversion of a shed into a strawbale dwelling. The shed is 12 metres by six metres by 3.5 metres high. The floor is concrete and the shed has two big sliding doors in the front. The owners had already acquired some salvaged doors and windows when they approached us.

Rudy suggested that they fit all the windows and doors to the existing structural building, but leave in place all the galvanised sheets which form the existing walls. The next step is to lay the bales butted up against the existing walls, using them as formwork. This way the bales are supported and kept straight, which is very important.

All the internal rendering can be done once the bales are up. The external sheets can then be removed from the windows and door areas. Providing all the necessary facilities, like the bathroom, the toilet, the laundry and the kitchen are completed and approved of by the Shire, the owners will be able to move in and complete the external rendering of the dwelling by removing the cladding sheets as required.

This method is also suitable for a number of other buildings including old asbestos-cement clad homes, providing they gain approval from the local council.

SIMPLEST METHOD

In our experience, the simplest way to go about strawbale construction is the post

The backhoe came in handy when it was time to do the rendering.

and beam method. This can be done by purchasing the steel and roofing required for the project and doing your own fabricating. Or you can purchase a modular prefabricated shed without the walls. We prefer strip footings between the uprights in preference to a concrete slab. An earth or stone floor can always be done at a later date giving an initial savings of thousands of dollars which can be better spent on building materials for the house.

Whichever way you go, once the structure is erected and the roof is on, you have a covered area in which to place the strawbales and to work. The Bale Up method is to tension chicken-wire along the length and breadth of the wall, attaching it to the steel uprights with tek screws. This forms a solid backing for the bales to be positioned against. Spray the footings with bitumen emulsion as a damp course. (We also find that this is an effective termite barrier).

Lay the first course of bales onto this while it is still sticky. Position trench mesh which is slightly narrower than the bales along the top of this course, secure to the uprights with tech screws or with steel tie wires, then cover the lot in a mortar bed and continue with the next course. This does away with the need for reobar to impale the bales onto as we have found it is often difficult to get a straight wall with this method.

Continue laying the bales, and do one or two more courses with the trench mesh and mortar mix. You then have a strong, solid wall. Spray the first course or two of the completed bales with bitumen emulsion — on an exposed weather wall the entire wall may be sprayed. This gives an effective damp-course and is easier to render over than black plastic. Encase the completed wall with chicken-wire, secure any loose sections with fencing wire and then proceed to render.

The completed house, which incorporates strawbale, stone and timber, has an area of 400 square metres over four levels.

GROWTH AREA

Strawbale construction is a fast growing building option in Toodyay and there are about ten strawbale projects going on at the moment. This is a very encouraging effort from a small country town with a population of just 3,800 people. In fact we call Toodyay the Strawbale Capital of Australia because it is situated in the middle of the wheat belt.

We are also fortunate to have very progressive building and health inspectors and the Toodyay Shire Council is people-oriented. We think that this is a contributing factor to the popularity of strawbale construction here.

For would-be owner-builders who have the vision and the dream, you are also going to need the backbone and the will, for the road to self-sufficiency is not always paved with smooth, even pebbles but also has rocks and craggy boulders. But even with the rocks and craggy boulders, it is well worth traversing to attain your goal.

We have now moved from our little humpy of the past six years into our beautiful new home which is not completely finished but it is comfortable and spacious, and rambles over nearly 400 square metres of rugged, rustic splendour on four levels.

We have had some very hot days and we've found the internal temperature to be at least ten degrees lower than that outside. We have also experienced a fierce bushfire which had flames almost licking at the door, a frightening experience but the house has stood the test and we came out of it unscathed.

Exactly two weeks to the day and the hour of the bushfire we were hit by a severe storm with 100 kmh winds, torrential rain and lightning and thunder. The force of it was awesome, and where other homes suffered damage and loss we still stand unscathed.

We could be forgiven for the thinking that the elements are out to test us but the story of the three little pigs looks more sus every day.

QUEENSLAND

Architectural philosophy leads to a beautiful home

"When I first mentioned strawbale to my wife, Dee, she said 'No way,'. Then I showed Dee *The Straw Bale House* and she was inspired like me." So says Ahtee Chia, the highly-regarded Vietnamese-Australian architect from Maleny in the Sunshine Coast hinterland who — with his family — built the second strawbale house in Queensland.

The house is 140 square metres of architectural marvel, from above resembling a bird with wings spread, and from the front showing off its superb passive solar design principles to catch the winter sun and exclude summer heat. In the middle of the house is a large open space under the colourful stained glass window specially-commissioned for the huge, sweeping clerestory window that makes the house so bright and airy.

The kitchen and living areas are central, the children's bedrooms and the bathroom are in one wing, and master bedroom and other rooms are in the other wing. The floors are all slate tiled, and the modern and Asian artworks and furniture blend beautifully with the earth-coloured strawbale walls.

Ahtee has researched many alternative building methods and he settled on strawbale "because it seems the most user-friendly. It requires the least technical knowledge, it's ecologically-sustainable, energy-efficient, and looks fantastic. Another big advantage of strawbale is the speed of construction".

The house took five months to build, with Ahtee and Dee working one month full-time, and the rest of the time on weekends. It took them three weeks to raise the walls, using 600 bales of wheat straw bought after phoning the Department of Primary Industries to discover who was harvesting where. Ahtee's main stipulation to the bale contractor was: "I want them extra tightly baled, and very dry".

Maleny gets lots of rain, and Ahtee casually remarks that they got over 700 mm in 24 hours one month. This influenced Ahtee's design: the house has a post and beam frame, partly so he could roof first, and keep the bales dry during construction. Ahtee used large ironbark posts for his frame, and had everything set up ready to go before hiring a large mobile crane for a couple of hundred dollars: "best investment you'll ever make on your house construction", he says.

The house also has very wide (one metre) eaves. "When we were building it rained all the time, but we just kept going — it didn't slow us up at all," he says. Nor did using recycled timbers throughout.

Maleny has mild to cool winters, and Ahtee says he and Dee are constantly caught out if they go out for the evening: they have to rush back inside for warmer clothes because they get used to the comfortable inside temperature — even with no heater on.

As an architect of 30 years experience Ahtee naturally had no problems getting his plans passed by council. He generously attributes this in part to the engineering specifications provided by Michael Kingsbury — another Maleny region

Ahtee designed his family home "for our easy-going, Australian lifestyle".

strawbale pioneer (see page 131).

Ahtee used aviary mesh (bird wire) to cover the bales in the walls, after deciding that chicken wire was too floppy for the job. Ahtee applied two coats of render. The first was made up of: 5 parts sand, 1 part cement and 1 part Plastermaster (a dry, commercial, lime plaster mix). His second coat was made from: 4 parts local (rich, red) dirt, 4 parts sand, 1 part cement, and 1 part Plastermaster. Ahtee's view is that in using an earth render with a small amount of cement he has avoided the extra labour of a full lime render. "Why bother limewashing? You'd have to redo it every few years," he says. Finally, Ahtee applied a commercial water repellent coat of clear 'Tech Dry', a product originally developed in Melbourne to protect mudbrick walls.

Architectural philosophy

Ahtee has one of the most logical and attractive philosophies for Australian homes I've ever heard.

"Because of our high rainfall, we can't build Santa Fe style homes like in the dry south-west of the USA: we need eaves. Also, culturally we are different: this is an Australian building. In an Australian building we can have free-flowing spaces from the inside to the outside to take advantage of outdoor living. In summer we spend lots of time outside on our verandah. And because Queensland winters aren't too cold, we're out there then too. You don't get that in many US buildings — this house is designed for our easy-going, Australian lifestyle."

"I wanted to express the natural characteristics of the building materials we've used. For example, natural earth colours (using the earth in fact!) and natural clear finishes on timbers. Essentially, I'm talking about an expression of Australians from the sea to the bush."

"The brilliant light is the next factor. In Australia this is a big contrast with many other countries, for example Britain. The house has clear lines contrasting with the natural curves of the strawbales. We didn't try to make the walls flat — we want the natural curves to be expressed."

"The building reflects natural resourcefulness and the materials are simple, just like Australians: we are resourceful and we

From above, Ahtee's house resembles a bird with wings spread, and from the front showing off its superb passive solar design principles.

live simply and unpretentiously."

HAND RENDER

Ahtee said that they started by using a trowel to apply the render, but threw the trowel away after five minutes, put on rubber gloves and rubbed the render on by hand. It was easier, they got a better finish, and it was quite a sensuous experience!

The other advantage of 'feeling' the render was that it was not applied too thickly: its maximum thickness is 25 mm.

Because they used a post and beam frame, Ahtee didn't bother with bale pins. All the bales are tied to the posts with fencing staples and baling twine, and then the bales are all tied to each other with baling twine. "I used timber bottom plates, which I dynabolted to the slab, but I now know from Michael Kingsbury that you can use special concrete nails to secure your netting to the slab".

Ahtee didn't like the idea of using dowels to pin his window frames into the wall bales. "I found they didn't stay in place — they went out of alignment, so I used timber studs on both sides of the windows, with the studs going from the bottom plate to the beam of the house frame. Then I could place the window exactly."

When Ahtee was finishing the walls, he stopped one bale from the ceiling, then filled the gap with plywood. For the ceiling insulation he used R2.5 polyester batts.

"When we built, we had lots of helpers. We advertised in the local papers saying that if you want to learn how to build in strawbale, come along to our building site to help and learn. We provided lunch — we couldn't work, but we supervised!"

And Ahtee's best advice to prospective strawbalers? "You don't have to know everything about strawbale to start. All you have to do is 'do it' and work things out as you go along." — *Alan*

From caravan to palace

A single parent with limited finances enjoyed the experience of building her strawbale house.

by Lottie
Sunshine Coast, Queensland.

TO SAVE money and also to get to know the land, I started off living in a caravan with my two children, who are now 11 and 14 years. I was always used to small housing and I didn't mind living in a cosy caravan, so I was in no big hurry to build my house. Because of this I was able to enjoy the whole experience of building, always slowing down when I needed a break.

I didn't mind the experience of making decisions on the spot. So many of them to make! I usually followed my intuition, but of course I also considered expenses and practicality. And, thank God, it has all turned out wonderfully.

Some experience

I took the project on as a single woman but I had some experience of building a house with my ex-husband, who is a registered builder. So I had some background and it didn't feel daunting for me. There was only a limited amount of money and after I did some costing, I decided to just start and see how far I could go. A local handyman helped me with the roof and I paid a local builder for occasional supervision.

I am lucky enough to live in a popular area and quite a lot of people come and visit this place. Many WWOOFers (Willing Workers On Organic Farms) want to stay to help on the land or to help with whatever needs to be done. I have an extra caravan for helpers to stay.

I met many interesting and beautiful people this way and I never felt alone with my project. People came up with interesting ideas that I hadn't thought of. Even the landscaping around the house was looked at and two little ponds were created with a waterway of rocks to the dam.

The bathroom is not in the house. Living in Queensland with so much sunshine, I want to be outside a lot. I know that one thing to make me go outside is having the bathroom outside. It's 20 metres from the house. Even while the winter evenings can be very cold, I am happy with that decision. There is only one bedroom in the house. My kids sleep in the caravan where they can receive their noisy friends and listen to their favourite music without disturbing me.

Walls go up

The walls went up in a weekend with help from about 60 friends and neighbours. It turned out to be a fun, social, work gathering, with lots of time for cuppas. There were probably too many people helping. Three people were constantly dividing the bales for infill. This is where we had to wait, because it didn't happen fast enough.

The first bales were put in place over the rods, which were stuck in the cement. The higher bales were secured with the insertion of bamboo sticks, these were about 1.5 metres long.

The only difficulty we encountered was

a wobbly wall where a window was placed. This wall was too long so we constructed some timber bracing from top to bottom to help stabilise it.

We took some time figuring out how to fasten the chicken-wire to the bottom of the wall. We started by securing the wire in place on the very top, using hooks made from wire. These hooks were first made from rather thick wire. But they were rather tiresome to cut because of the thickness. I bought thinner wire which was easier to pin into the straw but it bent as soon as it hit an obstruction.

After a lot of thinking and experimenting, we ended up tucking the chicken-wire between the ant-capping and the bottom of the bale. The ant-capping is visible on the outside and is hidden on the inside by pavers.

Once the chicken wire was fastened on both sides of the walls, we sewed the wire together with an improvised long needle. One person stands on one side, pushing the needle through, and another person on the other side receives it, pulls it out and pushes it in. We used baling twine for this. We did this with every second row, from the top to the bottom.

The niches were fun to make. The shape was decided and cut out with a chainsaw. The finishing off was done by pinning chicken-wire in place and filling the uneven parts up with straw. You can make any form you like with this.

Rendering

I found the rendering to be quite time consuming. I had borrowed a cement mixer in which we made up a batch of render and then poured it into buckets. We spread the mixture with gloved hands onto the walls. After a while we each discovered our own technique. For example, I grabbed a handful and slid it upwards, shaping it evenly with the heel of my hand. Others threw handfuls on the wall in clumps and then smoothed it out with the heels of their hands.

We all used a fine sponge to smooth it out more evenly or to wet the mixture if it started to dry out a bit. The rendering took the longest time. We used many pairs of gloves because they got torn by the chicken-wire. I decided to render with cement not mud because mudflies love my neighbour's mudbrick house and are slowly chewing holes in the walls.

The application of the render was in two layers and I decided to colour the walls with limewash. This can be put on with a broom, which is very easy. I was also told that you get a more even colour than if you only put colour in the last coat of render.

The rendering took a few weeks. It was funny when we came to the last part — everywhere else the walls were completely sealed. There were only a couple of metres to go until the top — the last chance for mice to escape — and did they ever! I had noticed there were some in the straw before but now they jumped out by the dozen.

The rendering of the outside wall allowed me to use my imagination. I put extra rendering on top of certain parts and I pushed broken tiles into the render for a special effect. I even cemented two mirrors in an outside wall for Feng Shui. The effect is lovely and some people call my home the 'gingerbread house'.

I wanted the inside to be as calm as possible so I didn't put things into the render. I also wanted to hang things up. I love the look of the walls, they give me the feeling of being in a palace. The house is especially beautiful in the evenings when the walls have such a mysterious, cosy, warm look.

North wall

I didn't use any strawbales at all on the north side because I wanted a house with a lot of light. So the whole front wall consists of glass with shaped, fixed windows on the top. The spaces above the doors and around the top windows were filled with the cheapest type of board. But now that I'm an expert at cement rendering I did this over the boards so it has a similar look to the walls.

Lottie says her house has turned out wonderfully.

Right: The walls went up in one weekend when many people came to help, and had some fun too.

Floor

I decided to have clay pavers on the floor. I put black plastic down as a moisture barrier and poured crusher dust over the top to lay the pavers in. Ants don't seem to like crusher dust. The builder cut the pavers, showed me how to level the floor, and how to lay the pavers. Two women WWOOFers and I finished the paving for the house and the verandah.

Living in the house

I am now living in the house. I haven't got the finances together yet for the ceiling and a proper kitchen. There are openings between the walls and the roof because I haven't put the ceiling in and I have covered these with material. Despite this, I can still feel the insulating effects of the strawbale walls.

I am so glad I decided to build with strawbale it feels so special, natural and beautiful.

Huge executive home in new Toowoomba suburb

Jill Disley had a simple brief for her architect: "I want a new home that's energy-efficient. I don't care what it's made of but it's got to be extremely elegant". Her architect designed a huge, green, 'U'-shaped strawbale home with outside and inside walls so straight — you can't tell the house is made of strawbale. The 'U'-shape of Jill's house encloses a paved swimming pool in the central courtyard. The house is set in a brand-new housing estate on the outskirts of Toowoomba, west of Brisbane, and the other houses have a similar modern, executive appearance.

When you walk inside it's obvious that the modern, angular exterior style has been carried through to the inside: the walls are standard, painted plasterboard, cornices and skirting boards: you can't discern any strawbale texture.

Some people might be surprised to see an innovative, alternative building material used this way but I think it's a good example of how versatile strawbale can be. Not everyone wants a textured, rustic appearance for their walls (I do!), and if strawbale is not confined just to alternative-minded owner builders, its ecological advantages will be taken up far more readily by mainstream home builders. — *Alan*

The house walls are standard, painted plasterboard, cornices and skirting boards: you can't discern any strawbale texture.

DIY metal frame resists termites

Building a strawbale house was a learning experience that left Peter and Laurel wanting to do it again.

by Laurel and Peter Edwards
Gatton, Queensland.

WE FIRST heard about the amazing concept of building with strawbales on the ABC's *Landline* program. We then purchased the excellent American book *The Straw Bale House*. We decided to give it a go and designed a small six by nine metre cabin which we planned to use for visitors on our 65 hectare property near Laidley, Queensland.

I am a sheetmetal worker and have a fully-equipped workshop on the property. I was able to fabricate all the various metal components for the building. We find metal far superior to wood as termites are a real problem in this area.

First step was the slab. I then placed ten millimetre steel spikes around the perimeter, before the concrete was too firm. These were 900 millimetres long and hooked into the slab and they were spaced the length of a bale. The straw was bought locally for about $2 a bale and it was very dry.

The next concern was how to 'lock up' the corners, so I made from 25 by 25 by 1.6 RHS corner pillars into which the bales could be slotted. The pillars were also critical in anchoring the roof. We had 200 millimetre Z purlins attached to the pillars and steel trusses from 50 by 50 by 2 RHS bolted on top.

We picked up the windows secondhand. They were timber framed and 900 millimetres by 900 millimetres. I made up heavy gauge sheet metal boxes. These were a big success because I could force them level after the bales were jammed around them.

Walls

I found the stacks of bales tended to be very unstable about two metres high (a good breeze would set them swaying). This was especially so when we started to put mud plaster on the inside. The weight of the plaster made it even more unstable, so I propped the wall with timber.

This instability was surprising because I'd thought that by anchoring the corners and hammering spikes through the bales all the way up, the wall would be solid. Anyway, I put timber along the top of the wall of bales and bolted and then squeezed them down with long bolts. However, there was still some movement so I paved right along the top with concrete. Pretty drastic but very successful — and it also provided a strong ledge for later use.

On reflection it would have been easier to lay the bales on the flat (wider) side thus increasing the stability. It would also have made the plastering easier because we would have had the 'sliced' ends of hay to push the mortar into rather than the long strands. I thought it was a good idea at the time as we'd increase the internal

Peter hammers spikes through the bales to help stabilise the walls.

area of the cabin. But I think it made things harder in the long run — next time I'll try the other way.

Rendering

We found the worst part by far was the rendering. I mixed up the mud in a concrete mixer with a little cement. It was a pretty sloppy mix then, with gloved hands, we went mad with mud. The first layer looked pretty crook but we just let it dry and put another layer on, then another.

It gradually looked better and better and it is very satisfying work. For the third layer I used a mortar board and steel float. We experimented with mixing various coloured clays from different parts of our block. We are very fortunate to have some rich red clay which made a beautiful mix. To get different consistencies we experimented with differing amounts of clay, sand, cement, lime and straw. The difficulty was in remembering what made up the various successful mixes. The weather also plays a large part in how quickly the render dries and whether it cracks.

We fitted out the kitchen, bathroom and toilet and put a verandah all around the perimeter. All materials were either on our land already or scrounged locally. We found building with strawbale to be very quick in the initial building stages but it takes ages to finish it off. The rendering of the walls seems to be an ongoing project. It's heavy work best undertaken by young, energetic people with lots of helpful friends.

At one stage our herd of goats (since sold) got into the yard and soon found that by butting the walls they could expose the bales — the rest is obvious — repairs needed. Only goats would be so enterprising.

Rewards

On the good side, we find the cabin has a wonderful, cosy atmosphere which is enormously appealing to us and to everyone who has visited it.

25 by 25 by 1.6 RHS pillars lock in the corners and anchor the roof.

The insulation is fantastic. It's cool in summer and warm in winter. We really love it, and for the last ten months we've had friends renting it so it's paying its way too.

Strawbales produce a more unusual, primitive effect which is very attractive. It looks like we'll be starting another strawbale house as soon as time and finances allow. Anyway, that's our story to date and we hope it encourages others to try out these interesting building systems. It's extremely rewarding to produce something that uses natural materials and can be made utilising products from your own block.

Strawbale in a hot climate

The framework of a steel garage was the starting point for a house extension that included a kitchen, storeroom and studio.

by Phillip Richards
Childers, Queensland.

WE THOUGHT our little house would be too small and having a kitchen in it would be too hot in our Queensland summers. We were right. The house stayed stifling hot in the evening but we wanted to use a slow-combustion stove to make preserves, cook our food, make bread, and heat hot water for those times when it would be overcast and we would have no solar power. Our answer was to have a kitchen annexe. We had read about strawbale building, and we wanted to explore it.

We bought a copy of *The Straw Bale House* book. It is a marvellous book and is essential reading, but we developed our own plans. Louise is a potter, so we had to build a studio for her large kiln. We also wanted a storeroom where all our produce could be kept cool. Finally we wanted a verandah.

We thought loadbearing walls might be beyond our skill, so we decided on strawbales as infill. The northern wall, which has the most sun, was to be made of straw for insulation. We wanted the storeroom section to be as cool as a cellar, so we planned on strawbale for all four sides and an insulated ceiling.

There were two final considerations. One, we were running out of money so the building had to be cheap, and two, we really wanted to get on with farming so we wanted a quick job.

FRAMEWORK

We ordered a steel garage without any sides and with enough modules for the size we wanted. This was six metres wide and 12 metres long with a verandah along one long side. This gave us a nominal kitchen area of six metres by six metres, a six metre by three metre storeroom opening off from the kitchen, and a six metre by three metre studio. The strawbales take up quite a bit of room so the actual room sizes are smaller.

The main part is on a concrete slab. We paved the verandah with seconds that we found at a special price. We made concrete pads that jutted out from the floor and put the combustion stove and the kiln on them.

Not all the walls are straw. Two are metal, being the surrounds of the kiln and the stove. Another couple of walls are window walls: we bought secondhand windows and doors and used these to make walls that were basically windows and glass doors in frames, with any in-between bits filled with structural ply.

We also built a frame wall where the sink is, because this wall had all the plumbing coming and going. I just didn't want to face the problem of cutting through the bales, although there are perfectly adequate solutions to that problem.

BALES

There were many kinds of hay in our area but there was no straw. We were almost stymied at the beginning and we cursed ourselves for assuming that straw would be

The north wall: the bottom course is wrapped in black plastic with strapping and chicken-wire used to hold the wall in place.

easy to get. In the end we found a produce merchant in Bundaberg who knew where to get straw.

Our straw came on a wet day, so it had to be unloaded from the truck and stacked under the roof. It would not be in the way but it would be convenient to retrieve. We found that we had two different types of bales: one had orange baling twine and the other had black. One lot was excellent and perfect for building, but the other tended to cause problems and was less satisfactory.

The strawbales had mice in them, basically, we walled them in. That was pretty Edgar Allan Poe-ish, but it was recommended in *The Straw Bale House* book. We stacked our bales carefully on the concrete floor. We put plastic beneath them in case water washed through and we did all we could to keep them dry and well-shaped.

The roof, floor, and posts were already in place because of the garage kit. This was cheap and quick and we put metal strapping between the posts which was pulled taut and screwed to the metal posts. These held the bales in place.

The bale wall

In the interests of economy and to get more wall from each bale, we put ours with the narrow side down so that the wall would be 30 centimetres thick and each course 45 centimetres high. This also saved room space. However, I don't think we would do it that way again because we found it very hard to poke wire through to tie the mesh onto the straw.

The concrete floor of the building had a damp-course, but I was worried that water might seep through the joint between the cement render on the walls and the bottom layer of bales. So we wrapped the first course in black plastic. It seemed a good idea. However, it also meant that we didn't have spikes anchoring the first course to the floor — contrary to most of the written material.

I tended to be a bit slap dash with the first wall. I tossed the bales into place and roughly straightened them. In later walls I realised it was important to go very

Phillip finishes the first coat of render on the studio wall. He found rendering gradually became easier and more fun.

slowly, test each bale, and plumb it accurately. This prevented sags, bends, bows and unevenness.

The bays (areas of wall between posts) were either three or six metres, and that meant that we had to make halfbales. Our first attempts were a disaster. Eventually Louise worked it out and would measure the size of the hole that needing filling, push some spare baling twine into the bale and then tie the two halfbales before cutting the original twine.

Stacking the walls was fun, dusty and tiring. The final and fifth course was the most difficult and really stretched our shoulders. Soon we'd done the easy part — a couple of walls.

HOLDING WALLS IN PLACE

To hold the walls in place we used horizontal metal straps that we erected between the steel uprights when we got the garage — we found these strong enough if they were pulled tight. On the inside we first put a layer of chicken-wire then we used several 50 millimetre by 25 millimetre wooden battens. Why this size? Because the junk shop had stacks of them! We positioned them vertically, spaced so that each bale was held by at least one batten.

The idea was then to pass stiff fencing wire through from the outside, wrap it around the batten and then push it back through so that the wire could then be twitched around the strapping on the outside, and the whole thing pulled tight. By and large this was successful.

The bales had extra stiffening with metal rods pushed down through them every two courses. For this we bought 12-metre lengths of steel rod and sawed them to about the length of two-and-a-half bales. I bent about ten centimetres of the steel to a 90° angle by sawing through it a little way and then bending it. This stopped the rod from slipping through.

There was one internal wall that did not have the strapping because it had a door in it. To stiffen this wall we used wooden dowels. Each was about a metre

The end: the wall on the left of the picture is strawbale and keeps the heat out.

long, and was sharpened at one end with a hatchet and rasp. There were three of these put through each side of the door frame and some more were put through the bales from the outside to pin them together.

Putting the battens up and tightening the fencing wire took quite a long time. Then we had to fix the chicken-wire. On the inside this went between the straw and the battens, and on the outside it went over the strapping. We used tie wire, and wired the chicken-wire to the bale twine. This was one of the benefits of having the bales laid the way we did with the twine running around horizontally.

Our strawbale walls have a gap between the top of them and the metal beam of the roof cross-piece, so we took the chicken-wire right over so that the top would be rendered too. Along the tops of these walls we placed timber wall plates and fixed these to the bales with metre-long lengths of threaded rod. (These are easy to obtain and cut to length with a hacksaw.) The straw held the threads tightly, and the rods passed through pre-drilled holes in the wall-plates that were screwed down with nuts and washers. These plates were screwed to the posts and main roof members for added strength. We will fix an insulated ceiling to the wall-plates around the top of the storeroom. This will help keep the room cool.

Again, getting the wire in place and tying it up took longer than anticipated. You need persistence and two people. Then we went around and stuffed gaps after the chicken-wire was on. This was pretty silly but we did not realise how important the surface preparation would be. This was an error that we corrected on the final walls.

Rendering

We hired the cement mixer for ten days. Those ten days were full of unremitting toil, mostly because we made errors and we did not know what we were doing. We found that the mix described in *The Straw Bale House* didn't quite suit us, although it gave us a start. We had to play with the mix of cement, sand and lime to get it right. We found that the best mix for the first two coats was a six-to-one ratio, that is, six shovels of sand, one shovel of lime, and one shovel of cement.

I began with a new, flat, steel trowel,

Leontine Oostra and Steve van Schooten have built a sunny, open plan nine by seven metre home with a separate bathroom and laundry — all overlooking their stunning bush slopes and dam at the Crystal Waters permaculture village in Queensland. Steve and Leontine used simple, exposed plantation pine roof trusses with opening vents at the top of the gable ends for summer ventilation. Their ochre-coloured concrete slab — and terracotta tile 'skirting boards' — contrast pleasantly with the limewashed yellow walls. They used only 175 bales of local barley straw, and built the whole house in only five months. — *Alan*

and couldn't get it to work. I used all sorts of scrapers, and even a machine that was supposed to spray the render on to the wall while you turned a handle. In the end I grabbed a handful and slapped it on. I don't say this was the best way, or even a good way, but in the end it was what we did. Our hands became very dry and we went through many pairs of rubber gloves to protect them.

At that stage we almost gave it away. However we persevered and gradually it became easier and more fun. It was a great joy to actually shape the wall with the cement render, to make curves around the windows, to make flat ledges, and to allow hollows to form. We loved it and enjoyed our free form house.

We worked around with the first covering, and with this we were simply aiming at filling in the mesh. The second coat was the main one. It flowed on more easily, and we used it to fill and shape. For the outside coat we mixed cement colouring, ferric oxide, which gave the walls a lovely, dull yellow look that resembled sandstone.

We painted the inside, the paint goes on well although the cement soaks up the priming coat. On the outside we painted over with a weak solution of bondcrete to seal out moisture.

Worthwhile

Although the job took longer than we had planned, we are very happy with the result. The building stays beautifully cool and we know it is fire-proof. We tried to burn a bale and found it almost impossible, even with a splash of kerosene.

In the three years living with our strawbale construction there have been no problems and we enjoy being in it. Nothing has changed or moved or performed in any way less than expected. The storeroom always feels cool in summer and remains about 22°C at all times. The walls give the feeling and appearance of massive solidity. It was cheap to build and it is aesthetically pleasing.

Suburban bliss with rock-solid walls

David and Heather Allworth have built a 20 square strawbale house in a suburban setting in Toowoomba that backs onto a park and the university, so Heather has a lovely walk to work.

"We used to live in a brick veneer house, with the garage on the north side . . . to keep the car warm . . . we hated it," says David. The Allworth house is a lovely 'Monet' yellow and the walls combine beautifully with the large, timber-framed glass doors and windows to give a bright, relaxed atmosphere.

The house has a cypress pine frame on a concrete slab, with studs to hold the windows in position. David bought 40 strawbales at a time from the local produce store. This was far easier than ordering all the bales then struggling to weatherproof them — it was also just as cheap. David and a helper built the walls and David believes that "any trained gorilla can build strawbale walls". The walls and rendering cost around $7,500 to $8,000.

To pin the bales David chose to run threaded rod on the outside of the bales, one every two metres or so, with a parallel rod on the other side of the bale, all along the wall. His top-plate — lengths of 15 mm-thick plywood, the width of the bales — could then be dropped over the threaded rods, and tightened down for compression. There was no sideways movement of the wall because it was secured on both sides — not up the middle. To secure the bottom, David simply drilled holes in the concrete slab, poked an end of the threaded rod in the hole, and used concrete glue.

There has been no cracking in the render. The local swimming pool builder did the spray rendering with a cement render supplied by the local concrete company to David's recipe of 8 parts sand, and 2 parts lime, to 1 part cement. The whole house was rendered in one afternoon for about $800!

The initial spray coat pumped the render deep into the straw and made it easy for the thinner, second and third coats to be applied by hand. David stuck to the same recipe but simply made it thinner and it's hard to imagine a more energy-efficient house for a suburban setting. — *Alan*

The Allworths' charming house is 20 squares with three bedrooms and a carport, and backs onto a park and the university.

Below: Jo salvaged pieces of concrete from the tip to set into mortar to make a highly attractive 'stone' verandah floor for her B&B cottage.

Jo's peaceful little cottage house cost between $16,000 and $19,000 to build.

Determined to recycle with style

Jo and Kerry Armstrong have a lush permaculture property in the hills behind Queensland's Gold Coast where Jo has built a lovely strawbale cottage overlooking a huge dam. She's also building a strawbale bed and breakfast cottage on their 6 hectare (15 acre) property.

Jo is one of the building world's resourceful scroungers and recyclers. She has a knack for using discarded building materials where they will look elegant and classy. For instance, a $10 glass door most people would ignore has been hung sideways as an unusual and attractive window. She spent hours of back-breaking work salvaging broken pieces of concrete from the tip to set into cement mortar to make a highly attractive 'stone' verandah floor for her cottage.

The walls of the strawbale cottage glow with a rich 'rhubarb' colour. The cottage is about nine metres long by seven metres wide, with a mezzanine bedroom. It has a treated pine frame and infill bales walls of sugar cane straw covered with bird wire. Jo used three coats of a 3:1sand:cement mixture for the render (with an ice cream container of lime per mix), then a 'watery' colour coat. The cottage shutters are made from an old dance floor and the windows are recycled. The floor is a coloured slab which looks like huge tiles.

All the fittings are recycled and the biggest costs were water, power, and the composting toilet. The house cost between $16,000 and $19,000 to build. Her best advice is: "Be a cunning recycler. That's the only way to do it." The house is furnished from recycled materials, right down to the lamps and cushions.

"Physically, rendering was the hardest part of the project, but a woman can definitely build a house like this," says Jo.

'The barn' is Jo's bed and breakfast project set around a central courtyard and Jo's builder-mate, Ross Bourne, has done a lot of the work here. The bedroom of the B&B is cosy and elegant, and Jo has embedded small pieces of old broken crockery in a line around the room, set in about 30 cm (12 inches) from the wall. Jo, of course, had to embed the crockery pieces while the concrete was still wet. The concretor knew this, "so he just did his deep breathing and muttered 'Go on. Do what you bloody have to do'."

Jo is the sort of person who is determined to do what she bloody wants to do, and do it stylishly and sustainably — her husband, Kerry, must have worked this out long ago. *— Alan*

NEW SOUTH WALES

Attention to detail gives delightful result

A lot of research, thoughtful design and experimentation went into this large strawbale house built in Sydney.

by Janet Dunn
Sydney, New South Wales.

WE BOUGHT just over two hectares at Galston and lived in the existing house for 12 months before building our new place, which I believe is the first complete strawbale house in Sydney.

We selected Simpson Wilson Architects, with Andrea Simpson as our particular architect. Andrea impressed us with her eagerness to work with clients who wanted to build an environmentally-friendly home that included the use of strawbale, recycled timber, and non-toxic products. Plans for our new house gained the approval of Hornsby Council. The council required no modifications nor were there onerous stipulations or requirements. Well done Hornsby Council!

DESIGN

The four main concepts that went into the design process for our house included: solar passive design; natural materials (no toxins); low embodied energy and sustainable construction.

The strawbale method we used was as infill for the walls. The design for a strawbale house is particularly important as the straw walls have such good insulation, that a poor design can result in a very hot house. The main house has been designed with the living spaces and kitchen on the northern side with windows spanning almost the entire northern face. The bedrooms, the bathrooms, the laundry and the pantry are on the southern side with far less glazing, but adequate for good natural light.

The eaves are large for two reasons: firstly to prevent too much rainfall hitting the walls which can cause damp problems with strawbales if the exposure is prolonged; secondly to block sunlight in summer, but not in winter. The floor on the northern side is a slab for thermal mass and the floor on the southern side is timber (raised a step up from the slab level) which will provide some ventilation but was mainly done for aesthetic reasons.

A pergola is situated on the northern side and will be covered with a deciduous vine to allow light and warmth in during winter, but it will block the sun in summer. A pergola has also been placed on the northern side of our games room which is a separate structure that is placed some distance from the main house.

The house design also includes loft space that can be utilised as a bedroom for our two sons while they are still young and it gives us additional flexibility for guest accommodation. There is also the outdoor room — this room faces north and is under the roofline but it has no northern or eastern wall. It is protected from the north and the west by the house itself, from the

The eastern wall of the bedroom wing: the earthen render was applied by hand and has a smooth finish.

east partially by the bedroom wing and it will have full sun penetration in the winter but it will be wonderfully shaded in the summer. A low strawbale wall will provide seating around three edges of the room.

Environmental factors

I have tried to use local sources wherever possible (although this has generally related to labour as most materials are made in factories all over the place and shipped to local shops), or products that require minimal energy in their production.

We have specified that all timbers are to be either recycled timbers or plantation timbers. We have purchased a large number of recycled doors and windows and try to use other recycled materials where the cost of doing so is not prohibitive. We have pulled down two sheds whose timbers have contributed significantly to the building process and we are yet to pull down a house which we will likewise recycle, if not here, then through other recycling outlets.

Further we have selected builders who are permaculturists. They share the ethic of sustainability and are happy to use and reuse recycled materials. This has proven to be a very important part of the process as they positively reinforce what we are trying to do rather than trying to put us down as I have found with some suppliers (a small minority).

Foundations

We dug down to bedrock (engineer's requirement), and had to remove a lot of sandstone floaters — they are absolutely beautiful, I just have to work out how best to use them! The footings were then formed and poured. For two of our buildings this consisted of strip footings 450 millimetres wide all around the perimeter of the building, and along any internal walls (one inside the garage). The footings are 450 millimetres wide to allow for the width of a bale. Then ant-capping went over the footings, with a space between ground level and the ant-capping in order to see if there is any termite activity.

The bottom-plates were laid down over the ant-capping and bolted into the concrete (creating holes in the ant-capping!). The bottom-plates are there in order to later tie the strawbales to the structure using wire and are simply two rows of wood laid flat on the footings, 450 millimetres apart.

The third building (our bedroom wing) has a timber floor which was constructed in basically the same way as a conventional timber floor. The footings for a strawbale home simply have to take into consideration the width of a bale versus the width of other conventional materials such as bricks. The only extra cost was that the ant-capping had to be custom-made as the widths we needed were not available over the counter.

The frame

The frame goes the width and height of the building but it must also go into the building the depth of a bale. A conventional frame usually only goes into the building by the size of the piece of timber. Our frame had to go into the building by 450 millimetres which reduces the internal space by over a metre once rendering is taken into account.

We minimised the amount of timber used in the wall frame by the use of ladder trusses or wall trusses. These are 'gang-nail' trusses manufactured to specification by a conventional truss manufacturer. The amount of framing is quite small: there are four corner posts and only nine trusses for a 6.5 metre by 10.5 metre building. This is significantly less timber than a conventional brick veneer home.

The wall trusses are all joined together by Hybeam, a laminated timber product made from plantation hoop pine with a high strength rating. It is used at the top of the wall trusses and the roofing rests on Hybeam which goes all the way around the building.

Roofing

The roof overhang is 700 millimetres which is a little larger than a conventional building. This provides the correct overhang for solar access in winter and summer and it provides the strawbales with greater protection from rain. Roofing consisted of putting the roof trusses up, then putting the fascia boards on (I found some beautiful recycled oregon fascia boards from a demolition yard in Brookvale).

The next step was putting on the guttering. We chose Smartflo guttering, but the plumber found it quite difficult to put up. However, the gutters do look good with a nice, slim profile but they still catch leaves and require cleaning. The cleaning process is much simpler though, and the leaves don't end up in the part that actually carries the water.

We opted for a galvanised iron roof which has turned out beautifully and will provide clean, drinking water. The roof has a very simple line, pitched at 26 degrees, and galvanised iron really suits the strawbale look.

The electrician and plumber came in and did their bits through the walls once the roof was on. Once again, there is not too much difference. We insisted on the electrician using steel conduit throughout any wall areas which prevents the conduit being accidentally penetrated by something sharp and creating a potential fire hazard within the straw. The plumbing is not really any different, we just had to ensure that there were pieces of timber placed in the appropriate spots so that the plumber could attach bits to them.

Walls

The buildings were then ready for the

An interior niche in the games room was cut into the strawbale wall before the rendering was done.

strawbale walls. Most of this was achieved during our wall-raising weekend. The first row of bales had to sit in, and on, the bottom-plates. This involved cutting a chink out of the bales on both sides so that they would sit on the ground properly. This was a lengthy process, and we have since decided not to do it this way. While these were being cut out, sections of high tensile wire were being cut, and holes were being drilled through the bottom-plates so that the bales could be tied into the structure once they were all up.

The second row of bales was much easier, as they only required shaping for the corners, or edges next to the wall trusses. The tricks to be learnt are that all bales should be laid with the knots in the baling twine uppermost. Also every bale has a 'cut' side, and a 'bent' side. Every second bale should be turned around so that the bales go 'cut — bent — cut — bent — cut'. This stabilises the wall. If this is not done, then the packing density of the bales results in the wall just slightly curving inwards towards the side that has the 'cut' edge.

Once the bales are up, they need to be bashed with a great big wooden mallet to get them roughly straight, and then they are tied down by threading the wire through the bottom-plates and up over the top, and then tightening them. This strengthens the wall far more than you would think. The wires are placed about 900 millimetres apart (about one bale length). The tightening is achieved by using gripples. These are fencing devices — just little things that allow movement of a piece of wire in only one direction — they are fairly expensive little things though, I paid about $600 just for these!

As you can imagine, halfbales are required in many spots. These are made by putting a special needle through a bale at the length of the required halfbale, and tieing another four baling twine ties (two for each half) after which you cut the original ties, resulting in two halfbales.

The wires that wrap the bales are also tied together through the bales. We had walls that were six and seven bales high. For this height, we tied each wire twice. Once at two bales high, and again at four bales high. These tie wires were tightened to bring the high tensile wires into the bales a bit more, so that they were not so prominent for the next step.

Once all was ready, the bales were then given a haircut with a whipper-snipper. This was a remarkably quick way to get the walls looking quite neat and tidy and ready for rendering. It is possible at this stage to cut niches out of the walls for all sorts of purposes. We have a niche in our toilet to hold the toilet paper, and to hold the essential toilet-time reading items. There are a few other niches in our buildings — mainly for decorative purposes.

Rendering

We decided on a mud-based render rather than concrete. Why didn't we choose concrete? We felt that earthen renders were simply nicer and we wanted to use natural products as much as possible. An earthen (or non-concrete) render will facilitate 'breathing' of the wall, and this is very important in a high humidity, sometimes torrential rainfall area, where the walls will suck in some moisture. It is absolutely essential that this moisture be just as readily sucked out again — using highly technical terms here!

The first stage of our building project comprises three separate buildings, and our builder convinced us to try different methods of rendering for all three buildings so that we could determine the best techniques and formulas for use in the main house.

We decided not to use any chickenwire around our strawbales. It appeared to be an unnecessary step, and therefore expense. All buildings had a minimum of three coats of render. The first coat was always a slurry or splash coat that was flung onto the walls using a large brush. This penetrated quite deeply (being such a wet mix) and provided the 'key' for the next coats.

The second coat was the thickest coat and fully covered the bales and got the walls fairly well straight. The final coat was meticulously applied to get a smooth and even finish on the walls. All the coats were applied by hand which was very labour intensive but the finish is an unbelievably professional look, with most people finding it difficult to believe that the render doesn't contain concrete as it is very solid and smooth — particularly the interior walls.

The earthen render was put on very thickly providing us with further insulation properties (not that we needed any more) and some thermal mass. I found a wonderful benefit of this thermal mass one very hot day — the walls were cool to touch and emanated cool air.

The garage

The render was made from clay, chopped straw and sand and was literally thrown at the wall. This technique may have worked well in the drier arid regions of the USA, but not here. The weather was very humid when we made this render and the straw began to decompose resulting in several different types of fungi growing — several of which are harmful to your health.

Fortunately this occurred in our garage, and so it wasn't considered a serious enough health issue to worry about pulling it all out and starting again. We switched to a lime render as the final coat in the garage, hoping that the anti-fungal properties of lime might help to alleviate these problems.

The games room

The interior walls were rendered using just clay and various sands, the exterior walls started with two clay coats, and then we put a beautiful lime render over the top, with a rough finish. On the south-facing wall we used a ten per cent concrete mix to see if this was going to strengthen the mix to provide the most weather-resistant finish for our most exposed wall. The other three walls were done with no concrete at all. The south wall was the only wall that showed significant cracking and so that was the last time that we used any concrete in the mix. The other three walls appeared to provide the same strength as the southern wall but lacked the cracking.

Clay renders and lime renders don't bond very well to each other and a bridge coat needs to be applied to ensure a good contact. Therefore, the third coat for the exterior of the games room was another slurry coat of a special mix containing casein (a milk-based protein that acts like a glue) to provide the necessary bond for the lime render. We also decided that we would put oxides in the renders so that we don't have to keep painting the exterior.

The bedroom wing

The exterior of the bedroom wing was done in a clay render using some cow manure as a hardener (it was pretty smelly while it was being applied but once dry there is absolutely no smell). The clay exterior is not as weather resistant as the lime. There are no real advantages to the use of clay over lime render except a marginal price difference — manure can generally be collected free, whereas lime must be paid for.

The interior of the bedroom has the most incredibly smooth finish, looking virtually like a perfectly smooth plaster wall, except it is all done with clay, sand, and some fine granite dust. It really is beautiful and attracts many comments. We chose a clay rather than lime finish for the interior of our living spaces for the health benefits of clay versus lime.

Next time

What will we do differently for the main house? Plenty!

- Firstly, we won't do further experiments but we'll capitalise on our experience. The interior will be a clay, sand and fine granite dust mix. The exterior will be lime render.
- We are investigating the use of a spray pump and operator to do the first and second coats on the house, both interior and exterior.
- We will spend more time ensuring a good line up of the bales at the wall construction time to minimise the rendering time.

Finished

We now have a completed games room which is currently our kitchen, diningroom, loungeroom and study. The completed bedroom wing is our bathroom, laundry and sleeping area. We traipse a distance of about 50 metres through mud and rubbish to get to our bedroom at night time (this is quite an adventure for the children). The garage is the builder's workshop and we use it for storage.

The main house is now under construction in between these three buildings and we are beginning to capitalise on the mistakes and lessons learnt from their construction. We are also experiencing the delights of living in a well-designed strawbale home — albeit a temporary one.

The Bower built of straw

Strawbale was the perfect material for a community centre that reused and repaired waste materials.

by Shane Naughton and Milosh Obradovic
Ewingar, New South Wales and Hardys Bay, New South Wales.

OUR INTRODUCTION to the world of strawbale building began when we saw a small flyer that invited people to join a reuse and repair cooperative for two dollars and eight hours of voluntary labour a year.

The cooperative began at an earthworks course run by South Sydney Council – as a waste reduction initiative. In time, a grant and a block of land were secured to build the facility which would house a delivery bay and a repair workshop where discarded material would be reused and repaired. Material which would otherwise end up at the tip.

The Environmental Protection Agency and the South Sydney Waste Board were the financing bodies involved. A coordinator was employed to work full-time and realise the vision on behalf of the membership. The site is part of an old army barracks and is where a workshop building burnt down around 30 years ago. The dimensions are 36 metres by 30 metres. A heap of soil had been dumped there. This was the first problem: we tried selling it as fill, with no luck. Then someone suggested making mudbricks with it but that was dismissed as too labour intensive.

We ended up scraping it to the back of the site and we discovered an old concrete slab. A few keen 'scavengers' set to work sourcing and collecting materials which could be used. They acquired eight, 12-metre span steel angle trusses from the demolition of a public hospital up the road. This dimension really determined the depth of the building. The rest was designed within the parameters of what was available — discarded material found on the side of the road, demolition Australian hardwood, scrap steel and industrial off-cuts.

The council drawings presented the design using the readily available, predetermined materials such as four by two hardwood trusses, plywood and corrugated steel. Finding enough material for the wall was not easy. We found an article in *Earth Garden* about strawbale building, its ease of construction, the low cost and that it's a waste product.

We rang John and Susan from Huff'n' Puff Constructions and organised a workshop within two weeks. John was also to come a week before the workshop to make sure we were on the right track. With the workshop organised we all set to work. Jane was the coordinator, Milosh the architect-builder, Ray and Shane the carpenters, Max the chief collector of materials, and Ben, Mim and Mandy were the backbone.

FOOTINGS

We wanted to use as little concrete as possible for the footings. To support the timber and steel frame of the building we used concrete block footings without steel rein-

The volunteers have a cuppa before stacking the bale walls. The foreground shows the brick footings with moisture barrier, ant-capping and hardwood base-plate.

forcement under each post. The wall is supported and raised off the ground with strip footings made of recycled bricks. Reusing bricks has a much lower embodied energy compared to using concrete.

Ten pallets of bricks were dumped on the site some years ago which seemed ideal until the day came to move them 300 metres, with the bricklayers due the next day.

We hired a large forklift that immediately got bogged. A four-wheel drive forklift was then called in, which also got bogged. We sacrificed a few bricks and built a road for it and managed to get three-quarters of the bricks out before getting the forklift totally stuck. What a nightmare! The rest of the bricks we moved the next morning using wheelbarrows.

The ten bricklayers steamrolled through the 600 millimetre wide, six course deep and 70 metre long strip footing in one day! Steel straps from a local scrap yard were set into the footings providing brackets for the posts. Once the footings were ready we laid a bituminous moisture barrier and galvanised, ant-capping, hardwood base-plate. Another method for doing strip footings if you don't have access to bricks is to use old car tyres filled with gravel or rammed earth.

Frame

The timber posts are 450 millimetres wide blades made up of four lengths of 100 by 50 millimetre Australian hardwood. These were picked up directly from the demolition site to avoid extra cost at demolition yards, which also greatly reduced transportation. Other frame materials included:

- Exterior plywood sourced as off-cuts from a desk factory; the off-cuts happened to be the perfect width for webbing.
- 100 by 20 millimetre hardwood battens from broken pallets — readily available, sectionally small, recycled hardwood.
- Factory off-cut plywood and noggings salvaged from old hardwood pallets make up the braced truss column.
- Strawbale walls provide the lateral bracing to the building. The ladder-truss composite parts were cut and assembled on site.

Walls

The base of the strawbale infill wall is wedged between the two hardwood rails which make up the base-plate. The small

The bales are compressed under a timber and plywood top-plate.

metre wide eaves protect the wall from the rain.

Roof

The 12-metre trusses and purlins were discovered and obtained directly from a demolition site. Moving the trusses was the first working bee. They really determined the width of the building. One-metre long, scrap steel angles were welded on each end for overhang extensions and two scrap steel straps were welded at each end to connect them to the posts.

Although the purlins already had connection holes in them, they had to be redrilled for the new spacing determined by the bale module. The purlins were not cut but laid in alternate directions to allow them to protrude further than the truss spacing.

gap under the straw improves the drying capabilities of the straw. A bale is 48 times the size of a brick and no mortar is used, therefore the initial construction time is quicker. The bales are compressed under a timber and plywood top-plate using high tensile wire looped around the wall. This is sufficient to prevent any sag in the bales over time as the wall is not loadbearing.

The bales are tied to the framing using chicken-wire. It is nailed to the framing and attached to the straw using wire prongs. The chicken-wire also becomes a substrate and reinforcement for the render. A lower cement render mix was used to allow some permeability of the wall, thus reducing condensation. One-

All steel for sheeting is recycled and was fortunately in very good condition. However, there is still the question of the old screw holes. When it rains the air on the outside cools relative to that in the ceiling cavity causing this warmer air to squirt under pressure, out through the old holes in the sheeting. This coupled with the fact that the holes are on the ridges rather than the valleys of the corrugation, prevents water penetration. Only a few bigger holes have been siliconed and the roof works. Strips of polycarbonate on the south provide natural light.

Floor

A smaller, existing concrete slab was discovered on the site. A combination of new concrete and 'freebie' pavers enlarge the old slab. The brick footings which raise the wall base one course higher than the finished floor level (the existing slab level), served as formwork for the additional strip of new concrete. Floor vents under each south door bring in cooler air at ground level through the grilles in the new concrete.

Windows and doors

By this stage there was quite a collection of assorted doors and windows. We built a box frame for all sides except where we could utilise the building posts. The 450 millimetre wide frames housing the salvaged window frames were made as C-shapes (L-shapes for doors) and always mounted against a structural post for a firmer fixing and a reduction in the amount of timber and plywood used. The strawbale wall easily accommodates building tolerances and different shaped and sized openings. High level windows on the north provide clerestory lighting to the space.

Like the windows all doors are fitted into tailored jambs. There are double doors on the south wall for improved thermal and especially acoustic performance. The inside set of doors all have some glass — western light reflects off the plywood reveal when the outside door is opened, providing additional light to the space.

Building

The time had arrived for the workshop. Everybody had put in 12 hour days for two solid weeks and we were still drilling the bottom-plate down and making window frames. To top it off the rain had started. Then the strawbales arrived looking like a hay shed on wheels.

Unloading the truck was the best part. Everyone got involved, even curious onlookers, as a truck full of strawbales is not a common sight in the middle of Sydney. By 9 am we had started unloading, and all 30 students for the workshop had arrived so by 11 am we had a nice hay stack in the centre of the building, the perfect place for a cuppa.

The size of the project now started to dawn on us. Had we bitten off more than we could chew? The students were now ready to build walls (the fun part). By this stage the rain had really set in and tarps were going up all over the place to protect the bales and give the carpenters a place to continue with the window frames. The rain did clear in the afternoon and within a couple of hours the first 20 by four metre wall was up, nine courses high, which was probably right on the limit.

Window frames were in and adjusted. John showed the versatility of straw walls and cut a square hole in the wall using a large chainsaw to place a window frame that was missed. We all learnt how to make halfbales, shaping bales, cutting in small alcoves, how to get a tight wall and numerous other tricks of the trade.

The next day the sun was shining and we were all feeling muscles we didn't know we had. By the end of the day the rest of the walls were up and were being compressed. Compression was done by threading high tensile wire through the top and bottom plates using the holes drilled earlier and wrapping the wall completely.

The two ends of the wire were joined up using gripples, a fence straining system, which was the quickest method. We made sure we alternated the gripples inside and out so the walls didn't bow.

Getting the top-plate level was a little tricky and having nine courses didn't help. We attached the top-plate ends to the posts with brackets. On the third day of the workshop we made adjustments to the walls using large wooden mallets.

Rendering

After the fine tuning, the chicken-wire was attached to the top-plate and stretched down to attach it to window frames, posts and the bottom-plate using nails. The chicken-wire acts as reinforcement for the cement render. The chicken-wire would have been unnecessary if we had opted for an earthen render, which is a must in the future.

Special pins were made to pin down any loose wire, this made rendering with

It took four months to build the Bower.

cement so much easier. We also made sure the window and door flashings were placed under the chicken-wire. If you are going to render over any timber, covering with tar paper helps stop the wood from swelling and cracking the render. By the time we were ready to render the roof trusses had been placed into position using a huge crane. The recycled corrugated iron was screwed in place.

John had given us a 45 minute lesson in rendering at the workshop, only now we were on our own and we were a bit nervous. We ordered sand, cement and lime and we hired a mixer and bought a few trowels. The first coat was the most time consuming, as you have to really slam it into the wall to make sure it binds with the straw. Fortunately we had plenty of volunteers. We used left over bales for scaffolding, working from the top to the bottom. The main thing with the first coat is to get all the straw covered. The second coat or scratch coat is much the same, making sure you have covered any exposed mesh and that hollows are filled in.

We made sure we scratched the render in between coats to aid adhesion. The coat with colour in it is the last coat, of course. We used a commercial pigment called sandstone which looks fantastic inside and out. Trying to get a half-decent trowel finish was hard and ended up being unnecessary because the end product looked a million dollars once we sponged it.

FINISHING

Once the rendering was done we could concentrate on the finishing touches such as gable ends, guttering, landscaping and those other jobs. The gable ends were made with a combination of recycled pallet timber as weatherboards and steel. The timber is prevalent, being a better thermal insulator.

Four months after we started the Bower Building was a reality. A marked temperature fall can be felt inside the building in summer and there is no need to use artificial lighting during the building's operating hours. It is one of the first strawbale buildings in Sydney and the largest of its type in Australia, which is not bad for a group of city slickers with very little money or building experience. Come and visit the Bower, at the Addison Road Community Centre, Addison Road, Marrickville, Sydney.

So natural that horses tried to eat it

The move was on for a house that was environmentally friendly. This strawbale home was built but rendering was done selectively while the owners started an organic olive grove.

by Julie and Malcolm Gilfillan
Araluen, New South Wales.

WE WERE in the mood to try something different. In Sydney we had built a big addition and done extensive renovations to an 1880s weatherboard farm cottage. It was all done in keeping with the style and feel of the old house.

Now, however, we were breaking free: of jobs, of the city, of most of our regular responsibilities, and we wanted to grow olives organically on our land on the Deua River. We did a permaculture course and felt comfortable with Bill Mollison's whole approach. So with thoughts of passive solar design, good thermal properties, getting closer to nature as well as the appeal of using a by-product of another industry, building with straw for our home in the bush won out. This decision was helped by my sister, who gave me a copy of *The Straw Bale House*. This is a really beautiful book and is filled with very practical information on how to construct with strawbales.

Structure

We had in place a concrete floored shed, 18 by 11 metres, with a corrugated iron roof that was supported by 16 bush poles. We chose the northern half of this shed for our house with the remainder to be used as a workshop. It was obvious that we would therefore choose infill rather than a loadbearing strawbale construction.

The next decision we had to make was where did we want each window and door? Malcolm, the real builder in this partnership, put in place uprights which all the windows and doors would be attached to once the strawbales were in position. All the timber, windows and doors were secondhand.

Bales

The 210 bales of wheat straw — all that could be fitted on the truck — arrived just before Easter, 1998. We were in the worst drought for 15 years. The river had stopped flowing and there was no grass, just dirt. Our 'townie' horses had to learn to browse on the river oaks. For them this load of straw was irresistible. After we'd unloaded and stacked the straw I put up an electric fence around the shed area and their hour of gluttony was over.

We bought the straw locally, through our rural supplier, at greater expense than we'd wanted. It cost $6.50 a bale, of which $1.50 per bale was delivery charge. However, the straw was lovely, very clean with no sign of mould and we were lucky to be able to store it under cover. On the off-chance that it did rain again, it was well protected.

Before we started building, we had the

The northern wall with rendered window-sills and wire over the straw.

chance to spend a day with friends on the coast. They had been to a strawbale workshop and had some good ideas that we adopted and others we chose not to pursue. We learnt how to split bales neatly without them bursting all over the place, and how useful a chainsaw can be for shaping the bales or cutting channels in them.

Our hosts had had their straw for some time, but no shed, so it was stored under tarps which had leaked and caused water damage so many of the bales had mould.

BUILDING

Had we planned to build in strawbale before putting up the shed, we may have elected to set rebar rods in the floor along the central line of the walls. The bales are placed over the rods then pushed down on them. More length can be added to the rods as the walls get higher, and they help give stability to the wall.

Malcolm, our problem solver, thought of an alternative. He made a base-plate of oregon, secured to the concrete floor with

A log and a car jack stopped the bow in the window frame.

dynabolts. These lengths of timber ran along the centre of the wall plan. A channel, located centrally between the twines, was cut into the underside of each bottom row bale, which then fitted snugly over the base-plate and was thus kept aligned in the wall plan.

We chose to lie the bales on their wider surface with the cut side facing inwards. It takes more bales this way and they use more floor space but we felt it would give a stronger wall. It was more secure, too. If the baling twine snapped at some future time then the bale couldn't go anywhere because it was jammed into position. It was easy to level up any pronounced unevenness in the wall surface at a later date.

The cut ends could be trimmed with a chainsaw, a brush-cutter, or my favourite, the arbortech attachment of the angle grinder. This was great for shaping the square edges of the bales, curving them in towards the window openings, for example. Using the arbortech was like giving straw a haircut. As you might expect, the bales were stacked like bricks. So every second row began with a halfbale.

Once the second row of bales was in place, we pinned it to the first row by hammering in rods cut to half-metre lengths with bolt cutters, two rods per bale. Our first panel, one without any complications like windows, went up quickly, but we were dismayed to find there was an alarming amount of sway in the walls. The vertical pinning didn't seem to help much against this sway.

We dismantled the wall down to the third row and Malcolm put on it a 4.5 metre length of oregon, lying flat and attached securely to the posts at each end of the panel. There's not much sideways flexibility in a flat-lying four by two was the thinking behind this idea.

The underside of the row four bales now had to have a channel cut in them, and then they were placed over row three with its plank. Pins, as usual, were driven through, avoiding the plank. We were pleased and impressed with the resultant lack of sway. We followed the same procedure after row six, and ended up with a good, sturdy wall.

Where the walls had windows or doors in them, this method of stabilising was not necessary. The framing had a steadying influence. Occasionally, the bales put too much pressure on a window frame, causing it to bow inwards. A log and car jack, set horizontally, helped reverse this pressure. After about three weeks of construction, I wrote in my journal: "It's looking good — like a thatched cottage — thatched everywhere but the roof, that is."

CHICKEN-WIRE ON THE WALLS

Chicken-wire is used on both the inside and outside of the walls to give added stability, to keep the straw bits tidily against the bales, and to help when we render the walls. We used about two and a half rolls of chicken-wire. Our ceilings are over two metres high, so we used two rounds of the wire and hung the upper round off the top beams with small fencing staples.

Holes were cut out around the windows and we left enough wire to wrap around sills which we attached to the frame with staples. Some patching was required around the window corners. Once the chicken-wire was attached, inside and out, we then stitched it to the bales. Malcolm made two pairs of smaller U-shaped needles.

We threaded tie wire through one hole, took it up along the length of the U, around and back through the other hole. Malcolm was on one side of the wall and I was on the other. We did two stitches per bale. This way the chicken-wire was held flat against the bale both inside and out.

I found it hard on my hands (any work with wire is), but it gave a much neater finish, and took away that 'hairy' look. An unexpected benefit was that if the horses got through the barricade they no longer had the satisfaction of grazing on our walls, and the wombat who had eaten out little bits from the bottom row of bales in several places, lost interest.

RENDER

We chose not to render the walls partly because of the extra work involved (we were expecting our first olive trees to arrive in August and still had heaps of preparation of the ground, fencing and irrigation ahead of us). But as well, we really like the look of the straw. It gives off its own special smell, too. An advantage of having the straw exposed is that it is easy to hang pictures and other things anywhere. A disadvantage is that there are always bits of straw around the place, despite the chicken-wire.

We did do some selective rendering. In the bathroom we've rendered up to shower level. Two coats of render and then a coat of paint with the consistency of lime putty. These walls are not at all even, and perhaps the render exaggerates this, but the look is quite pleasing, almost sensuous.

As well, we rendered the windowsills, inside and out, to make sure that water can't get down into the middle of the bales. We found that loose straw continually leaked out from the chicken-wire at the bottom. So we rendered around the bottom of the wall. Successful containment.

The lime putty for the render improves with standing. We got the recipe for the putty from *The Straw Bale House* book. One hundred kilos of hydrated lime (five bags at $5.50) were emptied into a 200-litre drum of water and mixed. Every day or so we gave it another stir.

When we were ready to render, we used one part of lime putty to five parts wombat sand (from their excavations down near the river), although any sand would do. We didn't use all the lime putty and found, a year later, when we were installing our slow combustion wood heater, that although the putty needed pounding and softening up, it was fine for rendering behind the heater and its flue.

WASTAGE

Malcolm has done a fair bit of building over the years and he was amazed at how little wastage there was building in straw compared with say, timber. Two bales broke when they were thrown off the

The north-east corner before the chicken-wire was put on, which put an end to snacking from passing animals.

truck, and we did lose one when the chainsaw accidentally snipped the twine when cutting a channel. That was the total waste. We had thought we'd need 230 to 250 bales, but we only had to buy two more than the 210 that were delivered. We don't have internal walls, only a corrugated iron privacy wall around the bathroom, so our bales were used for external walls only.

Comments

"It must be so dusty." Unlike hay, straw seems to be almost dust-free. We made up a big mattress of straw from the two broken bales and my brother and his family slept on it. Two of the boys are asthmatic and both remained very well during their stay.

Rain

"What happens when the straw gets wet?" As long as rain does not get in through the top of the walls it is not a problem. If the walls can breathe, any surface wetness quickly dries out.

Fire

The straw is densely packed, and although as a wall it breathes, it doesn't seem to contain great rushes of oxygen. It's not particularly successful if used as kindling on the campfire. We ignited the bathroom wall briefly while soldering copper pipes. It smouldered a bit, flickered, and died. If a wildfire came through we'd be no more vulnerable than anyone else.

Insulation

Excellent. Before we got our heater, on a –4°C morning outside, it would be 6°C to 8°C inside.

Vermin

We have rats in the ceiling (as we did in our weatherboard home). And they extract straw through the chicken-wire for their nests. It's good to take precautions at the tops of the walls, making sure the chicken-wire covers the tops of the bales inside and out where it meets the beams. Total rendering would deal with this effectively, too.

Our top ten strawbale buildings

Building a strawbale chook shed was the start of an exciting business helping to pioneer strawbale building in Australia for Susan and John.

by Susan Wingate-Pearse and John Glassford
Ganmain, New South Wales.

WE first learned about strawbale building when we read an article by Leo Newport. We both thought that we should give strawbales a go. We were building a dropped slab cabin in Berry when the client, Mary Knapp, asked if we would build her a chook shed.

The chook shed led to us building a privacy wall for our very good friends, Paul and Jane Collings, who wanted a pool fence. The pool fence used 450 wheat strawbales and we were hooked, we had so much fun building it and finishing the job with a built-in Balinese day bed.

So much has happened since then. There are several professional strawbale builders, architects, engineers and owner-builders who have contacted us, some who have attended our courses and many who have not, who are getting into the technology. We now have over 2000 names of interested people on our database and we are encouraged by the number of builders taking it up.

AUSTRALIAN PROJECTS

These strawbale projects are ones that we have visited and, in some cases, been involved with. Our top ten (not in any order) include:

1 Bill Mollison's cabin, circa 1987, in Tyalgum, New South Wales. As far as we can work out Bill introduced strawbale building into Australia. The cabin was built by Bill and his team, and there is a video of it available from the Permaculture Institute, Tyalgum.

2 Mary Knapp's chook shed built by Huff'n'Puff. This was the catalyst for our involvement in strawbale construction.

3 Harry Partridge's second storey extension in Killara, Sydney. This construction was very significant as this was the first approved strawbale residential premises built in Australia, as far as we know. Harry's house appeared on many television shows.

4 The house in Toowoomba built by Mark Liebke. This house was designed by architect Kathy Wallis for one of Mark's clients. It sits in a new subdivision and looks just like any other modern house but with a big difference — over 900 strawbales and green cement render. Mark has done a wonderful job on this building and it has received a lot of publicity. Mark's strawbale house is an example for those people who still want very large buildings for a modern family.

5 The Hansons' two storey studio and loadbearing stables, workshop and garage approved by the Wingecarribie Shire Council. Designed by Bill Duncan of Duncan

The Cask Shed, built by Bob Roberts of Huntington Wines, Mudgee, was the first strawbale commercial building to be built using jumbo-sized strawbales.

and Cole this 240 square metre series of many-coloured buildings was built by Huff'n'Puff and the Hansons.

6 The Buddhist lama's residence in Whyalla, South Australia. This was the first building to be built on the Whyalla Ecocity Site and was designed by Paul Downton from Ecopolis in Adelaide. It was built in eight days during a workshop conducted by Huff'n'Puff.

7 The Cask Shed built by Bob Roberts of Huntington Wines of Mudgee, New South Wales, the first strawbale commercial building to be built using jumbo strawbales. An absolute stunner.

8 The factory built by Martin Broadhead of Delpar Investments in Singleton and built from jumbo bales and a smooth rendered, cement finish. The way to go with industrial buildings: sound-proof and so cool.

9 The Bower, Marrickville, Sydney. A large strawbale infill warehouse and shopfront in the heart of one of Sydney's inner residential and industrial areas. Milosh and Jane plus the rest of this incredible team have built 280 square metres for just under $40,000, using nothing but recycled material and strawbales.

10 The next one does not matter who does it, or where, as long as it is done.

Innovations

We have been involved in a consulting role with two jumbo strawbale buildings built in New South Wales. The first one was built by Bob Roberts, managing director of Huntington Wines, Mudgee. He came to us with a request for information: how high can you go with loadbearing bales? The answer is: not high enough. Can we build infill with steel studs? Yes, we said, but it's not sustainable. Susan and I were out west and we saw the jumbo bricks sitting in a paddock, we immediately thought of Bob's problem and the answer was jumbo strawbales.

We were able to guesstimate that a strawbale would loadbear a corrugated iron roof with a 12 metre span. Extrapolation of the 5.6:1 ratio for width to height and 4.8 metres was the height. To cut a long story short Bob built the Huntington Export Cask Storage Shed with 100 jumbo strawbales (2.4 metres by 90 centimetres by 90 centimetres) and it worked. Bob had an engineer from Dubbo certify the loadbearing construction.

I hope that we can build all our factories and large commercial buildings this way. It makes so much sense, you can put the walls of a large area up in a day. The use of a cement render pump will see the render go on in a couple of days. The thickness of the walls achieves wonderful sound-proof properties and super insulation.

The factory at Singleton was another example where jumbo strawbales came into their own. Not all the commercial builders that I have spoken to about jumbo bales are convinced, however there are two good examples of what can be done, and to a budget, already standing in New South Wales.

CARBON CREDITS

It just makes pure commonsense that if we can sequestrate carbon into strawbales and then use them to build with we then should be able to obtain credits for doing so.

We need to find out how much carbon, an acre or a tonne of wheat straw, for instance, that we can store in the form of a building block (strawbale). I have been advised that the figure of carbon stored should be somewhere near the same amount as a tonne of timber can store. For example, a fast growing eucalypt plantation averaging a stem growth rate of 20 cubic metres of wood per hectare may yield 500 kilos of dry wood per cubic metre (ten tonnes per hectare) and contain 50 per cent carbon (five tonnes per hectare) in one year. Source: State Forests of New South Wales.

If straw has the same amount of fixed carbon per weight as wood then we can do some interesting extrapolations. Rice and wheat cereal crops yield five tonnes of waste straw per hectare. In New South Wales we burn 600,000 to 2,000,000 tonnes of rice straw every year. If we take the figure of 600,000 tonnes of rice straw at a ten per cent moisture level then that means we have 560,000 tonnes of dry rice straw available for building.

Now let us assume that by using the 560,000 tonnes to build houses then we have enough rice straw to build 45,000 by 150 square metre houses a year. Not only that but we have sequestered 280,000 tonnes of carbon per year or six tonnes of carbon per house.

THE BUILDING CODE OF AUSTRALIA

In 1996 the Building Codes Board of Australia changed the BCA from being a prescriptive code to a code based on performance. This has meant that innovative building technology can be practised in Australia provided that the method deems to perform to the standards set by the BCA.

Strawbale building can be used if it performs to the requirements of the BCA. To give you an example of what it has meant to the average punter: in Adelaide an architect, Paul Downton of Ecopolis, can have his strawbale house plans approved in four days by a qualified certifying engineer without going near the relevant council. And the cost for doing that is 75 per cent of the normal council fees. The engineer now does the work of the council building inspectors. Only four years ago our council said that they would not approve a chook shed until we had the technology incorporated in the BCA. That was in the days of a prescriptive code.

Modest appearance gradually reveals stunning character

The dramatic curving lines of the clerestory window bring to mind the sweeping sails of the Sydney Opera House.

"We wanted to create the inside atmosphere of a friendly English pub," says an extremely modest Neil Cuthbert about the artistic and welcoming atmosphere of his family home near Orange in western New South Wales. Neil and his wife, Jacqui, have quietly gone about building one of the most impressive homes you will ever see. From a distance it's deceptive: you can't appreciate the dramatic curving lines of the clerestory window which brings to mind the sweeping sails of the Sydney Opera House.

You don't realise until you're inside that every room has charming views of the countryside, or that the passive solar design means natural light and warmth help heat every space. Even the spreading, flared shape of the external walls, bringing to mind a Tibetan monastery, isn't immediately apparent. In fact, this house crept up on me, and came back in my mind to surprise me long after I'd left it. It's one of my favourite strawbale houses in Australia, and it's the main picture on the front cover!

Neil and Jacqui's 16 square house was designed and built by Steve Sainsbury, and they speak very highly of his skills and application. Steve lived on site in a caravan and Neil and Jacqui were the labourers, organising materials and working bees. They started building in September 1998 and finished in May 1999. The house has its open plan living room and kitchen in the middle, with a wing on either side for bedrooms, studio, and a bathroom on the south side that has a full-length window opening into a fernery. Through their first year the house stayed a steady 22°C despite frequent sub-zero, frosty days in winter, and summer temperatures reaching into the high thirties. No doubt the double glazed windows and lambswool batt insulation help with summer cooling and winter warmth.

The bale raising took only two weekends with five or six extra people to help. This contrasted with the 13 days of hard heavy work it took to build the rammed earth internal walls. The massive internal walls are excellent for storing the day's heat to radiate through the house after dark (but you can't help comparing the effort against building strawbale walls).

The Cuthberts covered their strawbale

The Cuthbert house seems inspired by Japanese philosophy, and every room has charming views of the countryside.

walls with a wire product called Renderlock, which is specially designed to take render: because of the angled 'fin' of the wire, it won't float to the surface after an initial render coat, like chicken wire. They laid the strawbales straight onto the slab, and every second course they hammered pins vertically into the bales to overlap with the pins underneath. Neil and Jacqui said that the most time-consuming part was cutting and retying short bales to fit into odd corners. They used 500 bales of local wheaten straw which cost $2 a bale plus $1 per bale for delivery.

They made staples from fencing wire, and applied three coats of render. Their render mix was: 8 parts sand, 1 part cement, and 1 part lime, and they put a bit of sugar into the mix to act as a plasticiser. They also soaked the lime for two days before using it.

"It was great fun putting on the render because you could actually see a wall going up," says Neil.

The Cuthberts have used only recycled or plantation timber throughout the house. They also used Ortech strawboard panels which are a great alternative to standard plasterboard panelling. Apart from the environmental benefits of using recycled timbers they found a cost advantage: new oregon 25 X 7.5 cm (10 X 3 inch) beams cost $22 but the recycled equivalent cost only $11 each. The Cuthberts have a massive eucalypt post in the middle of the living room — hewn from a dead tree in a paddock — the post has large beams 'tied' to it to provide the large open plan living space.

The house sits back into the land to protect it from the prevailing southerly winds, and faces the north to soak up the winter sun. Some people have told Neil they thought the design was Spanish-influenced, but it seems more inspired by Japanese philosophy. Neil talks about 'kura' — the soul of the house — being around the kitchen, living room and fireplace.

The rear, south wall of the house, behind the main corridor, is lined with storage cupboards, but, cleverly, has narrow windows running along at head height so the storage areas are naturally-lit and not poky, forgotten corners. The fireplaces in the living room and one end room are built using old kiln bricks in a rounded beehive shape.

The bathroom is something special: the huge sunken bath has a shower attached to the wall with just two coats of Bondcrete straight onto the bathroom wall to protect the strawbale from the shower moisture. The full length window opening onto a private courtyard makes for a surprisingly sunny south-facing bathroom. And the slatted bathroom bench, which looks as though it's made from the finest, most expensive cedar, is in fact cypress salvaged from the remains of an old dog kennel sitting in a paddock. Very clever indeed.

"The idea of the house is to go through narrow spaces to get to open areas — it's a house of contrasting feelings. We like the idea of lots of roof and low eaves, to give the house character," says Neil. The Cuthberts have succeeded admirably. — *Alan*

Champagne views on a clever budget

Sam's golden rule was to collect his secondhand materials first, then design the house to fit the materials.

Sam Statham is building a large strawbale home on a gentle slope surrounded by grape vines on his parents' organic farm near Canowindra in western NSW. Sam's house has a post and beam frame of cypress pine on a slab, with an area of 185 square metres plus the mezzanine floor. Sam and his family and friends have all pitched in, so the house is a real celebration of fun working bees, and community and family participation.

The house used 450 bales of local straw and Sam found that the best way to even off the edges was with hedge cutters. He built an extension to the farm tractor's forklifts to lift the heavy posts and beams into place.

A highlight of Sam's house is the Japanese-style glass-walled garden on the eastern side: morning sun and warmth will pour in through the filtered green light of his inside garden!

Sam has used secondhand materials throughout, and he expects the final cost to be $35,000. His golden rule was to collect his secondhand materials first, then design the house to fit the materials. So far, Sam's costs are:

Windows and doors	$2,000
Concrete slab	$8,000
Poles and transport	$3,900
Roof timbers (3 X 2 inch)	$1,000
Other timbers (6 X 1 inch)	$1,000
R3.5 golden fleece insulation	$2,000
Hessian and roofing iron	$2,000
Strawbales	$1,200
Bathroom fittings	$ 500
Plumber	$ 150
Septic tank	$ 500
Dowmus composting toilet	$3,000
Electrician	$ 500
Sundries	$1,000
Running total	$26,750

Sam is happy with his earth and chaff render — he added sand to the clayey soil to 'customise' his render. He found it just as fast to mix the render by hand as using a concrete mixer. It took 15 minutes to mix one bathtub — enough to render a section of wall about 2.4 metres (8 feet) high X about 2 metres (6.5 feet) wide. After two earth and chaff renders he applied a coat of lime render. The red soil in Sam's render gives the house an air of lush fertility that reflects its site: grand views over a fertile valley of organic grape vines. — *Alan*

Heavenly work space a strawbale showcase

"Excuse me, what time is the church service please?" an elderly couple earnestly asked Margo Stephens last Easter as she worked away in her new strawbale studio on a hill above Mudgee's spreading vineyards. Understandable, really: Margo's studio looks for all the world like an old country church, with its gothic windows, its soaring, high-pitched roof, and medieval wall finish.

The studio is about 9 metres long and about 5.5 metres wide, and the ridge beam of the roof is 8 metres high. Margo and her late father travelled about western New South Wales measuring the dimensions of church windows and roof angles and the result is a startlingly sculptural and peaceful work space — perfect for a busy young sculptor. Margo's husband, Paul Phillis, casts sculptures in bronze: "She hands them to me and I turn them into bronze and silver". The pointed gothic windows come from Paul's careful craftsmanship — he steam-bent the cedar timbers on a jig he built himself then set them in a rectangular window frame pinned with dowels into the strawbale walls.

Margo and Paul started with a pole frame made of scribbly gum from Duneedoo (sounds like a line from a Banjo Paterson poem). And then the gleaming golden rafters came from cypress pine cut from the bush on their property. They borrowed scaffolding from a nearby winery and rigged up a pulley system to swing the strawbales high into the gable ends.

"My girlfriends helped me render the inside in two days — we had lots of working bees," says Margo. Margo and Paul brought in 300 bales and had about 25 left over. They 'impaled' the bales on 1 metre-long reo rods sticking out of the concrete strip footings. They also came up with a clever method to damp-proof the window sills and corners: they used sarking (paper-backed foil) wrapped around the corners and sills, so there's no chance of the rain sitting on the window sill, or worse, running back into the wall.

Margo used the local concrete water tank maker to spray render the outside of the studio. "He had a silly grin on his face the whole time, like he was thinking 'You mad people'," says Margo. Yellow and red ochre mixed into the render give the studio its lovely colour.

At the front of the church Margo used 150 year old roofing shingles which were salvaged from under the roof of the old convent in Mudgee — how fitting that the shingles have helped reroof a churchlike structure. The floor inside the studio is fascinating: it's made of rounds of yellow box set into a bed of sand. A big old yellow box had been knocked down and Margo and Paul saved it from being burnt. Once the rounds dry, they will sand and finish the timber.

WHY DID THEY CHOOSE STRAWBALE?

"I looked at a $3 bale and it would have been at least five big mudbricks to cover the same area," says Margo.

The studio is a perfect space for special ceremonies: a year ago Margo and Paul had

Above: The 9 by 5.5 metre studio cost around $15,000 to build — including the cost of a local builder.

Right: Margo and Paul had their wedding ceremony in Margo's studio, followed by a candlelit dinner with two guitarists playing.

their wedding ceremony there, followed by a candlelit dinner with two guitarists playing. The studio cost around $15,000 including the cost of a local builder, so Margo has gained an inspiring and attractive work space that's also a great example of successful strawbale building. — *Alan*

The straw-coloured walls of this home at Leura give it a vibrant feel.

Mountain-style home perfect for climate

Wayne and Fiona Gipters took about a year to build their superb mountain-style home at Leura in the Blue Mountains.

They decided to build in strawbale because they wanted high energy efficiency in the cold mountain climate. Also, their relative — and the builder — Scott McGilchrist, was very enthusiastic about strawbale.

"We thought it was something we could do ourselves," says Fiona. The house has a post and beam timber frame, with the strawbales on bottom-plates fixed to Besser blocks on strip footings. The timber floors are a major feature of the home.

The straw-coloured walls give the house a vibrant feel — as you can see on the back cover of the book! The unusual design means both bedrooms are on the mezzanine level. They're both very large, with one running the length of the house, and both look to the front. There are no internal doors except for the bathroom, so there is a large, open-plan living area.

"We built the house for luxury, and that's what we wanted. We love the feel of the house. As soon as you walk in the door it has a special feel — it would have been hard to achieve all the curves with other building materials." says Fiona. Wayne and Fiona positioned the house to avoid knocking over large trees. As one of the first completed strawbale houses in the Blue Mountains, and the first (of a growing number!) of strawbale houses built by Scott McGilchrist, the Leura mountain home is something very classy and very special. — *Alan*

Quiet elegance on Blue Mountains suburban block

Before Katherine Perrin designed her home in the Blue Mountains, she read many books on passive solar design. "So I made the house a strawbale rectangle with an attic on top for simplicity and energy efficiency," she says. This 160 square metre, two bedroom home could be in a French provincial town.

The house has double-glazed French glass doors on the north side to let in the light and heat, to warm the thermal mass of poured earth internal walls and suspended concrete slab. The ceiling is insulated with fibreglass and when it snowed this year, the house was 10°C with no heating and no window coverings.

Katherine's (experienced) builder was a first-time strawbaler. The house has a steel post and beam frame, and the infill walls were covered with chicken wire before the render was sprayed on. The sand in the render created the muted sandy colour of the walls — no oxides were added. The understated elegance of the wall colouring adds to the charming and harmonious air of the house. — *Alan*

Katherine's two bedroom home is a model of energy efficiency.

The Buntons' 10 by 13 metre house in the Blue Mountains.

Creative mountain home fires imagination

Matthew and Fiona Bunton were thinking of building in mudbrick on their 0.6 hectare (1.5 acre) bush block in the Blue Mountains, but when their architect, Rick Mitchell, visited with a trailerload of strawbales for his permaculture garden, Matthew casually asked him how much he knew about strawbale building.

"Since then we've never looked back," says Matthew. Their four bedroom home reminded me of wayside inns on the trekking trails between Nepal and Tibet, and this was indeed a design factor. The house has large, north-facing windows and glass doors opening onto a patio, with mudbrick internal walls for thermal mass, and recycled materials used wherever possible. Matthew and Fiona are owner-building the house on weekends and holidays.

"Owner-building is a big time commitment but we're not on a deadline, and we have Rick's expert advice," says Matthew.

The first coat of render on the walls was machine-sprayed with a render mix of one part cement, four parts sand, four parts loam and one part lime. The second, thinner coat was applied by hand with trowels, and they applied the final coat — "thin like paint" — with brooms. The third coat was one part lime and one part cement, with Bondcrete added, and eight parts local earth for an ochre-brown colour.

The site is surrounded by banksias with thick trunks, and we'd referred to the home as the "Bad Banksia Men house"; maybe "Namaste" the lovely, polite Nepali greeting, is more appropriate. — *Alan*

One day a machinery shed, next day a strawbale home

Jake turned the farm machinery shed into a beautiful four bedroom, 16 square home.

In 1992 Jake Jarrett bought a property near Orange and set about establishing a vineyard. The farm's machinery shed was about 20 metres long and about 8 metres wide, and one day Jake decided to convert it into a strawbale home. He kept the roof, steel frame and concrete slab floor, and bolted steel bottom-plates to the slab.

These held the steel rod that he used to compress the infill bales, and hey presto, he ended up with a beautiful four bedroom home. You'd swear the house had always been there.

The house looks down a slope over grapes that thrive in the rich soil. Winters are cold here: lots of frosts and about three snowfalls a year, so it's perfect for strawbale. A small electric heater, put on in the evening for a couple of hours, is all Jake needs to heat the entire house.

The open plan kitchen/dining/living area combines nicely to create a very impressive home. The lemon yellow exterior walls are bright and happy, and the fragrant jasmine creepers add a delicate touch to the timber pergola on the north side.

Jake had some help from his neighbours, son Justin and daughter-in-law Pip, and the whole shed conversion is a stunning example of a clever way to produce attractive, eco-friendly shelter. — *Alan*

The meditation centre is built with a heated, suspended slab among superb Blue Mountains native gardens.

Perfect contemplation

The Vipassana Meditation Centre is set in beautiful native gardens on the outskirts of Blackheath in the Blue Mountains. It's easy to see why meditation students are attracted to the centre: the gardens and the whole centre have a peaceful, relaxing atmosphere.

The centre has built a large strawbale building with teachers' residences either side of a large hall used for 30 day meditation courses (the centre also runs much shorter courses).

Dean Monahan, a local alternative building enthusiast, helped build the hall which is very carefully sited close to existing gum trees — it looks as though virtually no vegetation was disturbed to build the hall. The building is interesting because it has brick foundation walls on three sides to bring the floor level with the upper slope. Then there is a suspended concrete slab resting on the brick foundation footings. The slab is heated, so people meditating are warm and cosy! The interior walls of the hall have been rendered with a lovely, rich, cream colour (see the front cover of the book). Judith and I left feeling that if we ever do a meditation course, this is the place to go. — *Alan*

- ***Vipassana Meditation Centre: PO Box 103 Blackheath, NSW, 2785.***

Pioneers do house and land for $20,000

John Glassford and Susan Wingate-Pearse are well-known through their business, Huff'n'Puff Constructions, and in May 2000 they were building a loadbearing house with recycled materials and innovative, cost-saving ideas in the NSW town of Ganmain.

"Where else can you buy a block of land for less than $5000 that has town water, gas, and sewerage? Ganmain is equidistant between Melbourne, Sydney and Adelaide, and we hope the town will become a centre for sustainable building," says John.

John and Susan are using this project for the workshops they hold four times a year. People stay at the old Royal Hotel, and learn low-cost, owner-built construction methods. The focus of the workshops is 'how to build sustainably and cheaply'.

Low-cost methods include using rubble footings rather than full concrete. The loose rubble concrete in the trenches is 600 mm deep, with 75 mm (3 inch) stones in the bottom. On top is a 15 cm (6 inch) concrete bond beam. This method used only 22 bags of cement, rather than the 32 cubic metres of concrete in a 'normal house'.

The house is about 960 square metres, and on top of the rubble strip footings sits tin ant capping the width of the footing to act as a damp-proof course and termite barrier. Above this are bottom-plates made of 10 X 5 cm (4 X 2 inch) native cypress pine, dynabolted to the footings.

John and Susan are building a loadbearing house with recycled materials for less than $20,000 — including land and services such as gas, sewerage and town water.

Pavilion-style home with a holiday atmosphere

Justin and Meg Byrnes and their children live in a strawbale house near Orange in New South Wales that looks over a superb valley of native bush, and has a contrasting view of vineyards behind the house. Their home is built on a flat site excavated from a slope. It is a single-storey building with gable-end windows, a conventional concrete slab floor, and timber frame. The walls were cement-rendered by hand with an orange oxide colouring in the final coat.

Internally, the walls have been rendered but not around the windows, which are lined with sheets of pine plywood, as are the window sills. The timber window frames are intriguing: half the window is glass, and the other half is flywire with a galvanised iron shutter, which can be shut to cover the flywire in cold weather. The open plan living, kitchen and dining room areas all face north through lots of glass, opening onto a large terrace. These interesting features give the house the feel of a relaxed, pavilion-style holiday home — it almost feels like a classy beach house! The solar hot water service tops off a very environmentally-friendly and energy-efficient home. — *Alan*

Then they have laid the wall bales, covered them with wire, and used fencing wire attached to the bottom-plate to compress the bales, holding the wire tight with medium-sized fencing gripples. They do three or four compressions, getting about 5 cm (2 inches) of compression the first time, and about 7.5 cm (3 inches) of compression altogether.

The window frames 'float' in the strawbale walls but after the final compression they are rendered in place. John and Susan plan to do two coats of earth and chaff render followed by a final coat of lime putty. They have used recycled windows and doors, roofing iron, and other cost-saving measures for a staggering budget: the whole project — including land — will cost $20,000.

"We're concerned about the direction of Australia's society and economy — people need affordable housing," says John.

John and Susan have experimented with earth and chaff renders. The chaff is mixed with the mud and forms a cheap, strong, crack-free finish. They rendered a wall of Jack's Flat — an ancient building they've relocated — with the earth and chaff render, and after nine months with no treatment, it's as good as the day it was applied. — *Alan*

- ***For workshop details see the list of contacts on page 154.***

SOUTH AUSTRALIA

First contract-built home in South Australia

Josephine is delighted with her home designed around an atrium.

by Josephine James
Mount Barker, South Australia.

IN 1993, I bought ten hectares of land at Wistow and then thought long and hard about the sort of house I would build. It was a very windy area but I had to take advantage of the marvellous view from the top of the ridge so by placing the house corner-on, the prevailing wind would be split as the bow of a ship cleaves the waves. I decided that the house should be built around an atrium to take advantage of the warmth of the sun in winter and yet be protected from the wind.

I wanted to grow lots of native shrubs and trees to be able to sell cut flowers so to suit these surroundings I felt the house would need to snuggle into the landscape. Also, I wanted to utilise natural resources as much as possible and I had a lot of ideas such as solar-heated water flowing through pipes under the floor in winter and solar and wind power generation.

Sad to say, the technology is there but it's not easily affordable. Solar heated water under the floor was prohibitive and so was alternative energy. To have home-generated power flowing back into the grid requires a very expensive installation in country South Australia. In fact, I do not believe such a facility is yet available in the urban areas. Why? I cannot imagine, because we'll run out of oil one day.

WHY STRAWBALE?

The land contains a great deal of rock and a stone house would be marvellous, but someone has to collect the rocks at a price. A friend suggested strawbales as great insulation in both winter and summer, but I was unsure about this. Would strawbales be strong enough? Would there be a fire problem? Would there be a vermin or termite problem? And so on. However, I did consider mudbrick and rammed earth.

After I had worked out a plan that incorporated all my ideas, I took it to an architect and he also asked if I would consider strawbales. I said: "I've heard of this, so tell me about it." He lent me a book on the subject and before I was halfway through I was convinced that strawbales were not only a good building material that would provide excellent insulation but using them would be a responsible way to build. There was no problem with strength, there was less fire danger than many conventional building materials, and the same applied to vermin and termites.

Straw is a byproduct of wheat-growing. It is produced every year and grows in six months. In other words, it is sustainable. The earliest that pine wood can be ready for milling is eight years after planting.

Straw is often burnt if the farmer doesn't need it, which adds to air pollution. Sometimes it is ploughed back into the land, which is a better option. By us-

Above: Bales being compressed with car jacks placed on top of framed wall.
Right: Josephine lays the first strawbale.

ing straw for building, we put a waste product to good use and save some timber, or some holes in the ground for brick-clay, or some holes in cliffs or hills for stone.

So straw it was! But we had to compromise because of cost. Strawbales take up a good deal of space on interior walls and I was not prepared to settle for small rooms so it was decided to build the outside walls of straw and the interior walls of stud and plaster, with the exception of the atrium which would all be of glass.

Although I am pretty active for my age, 69, I decided that owner-building was not for me. I wanted it finished in the shortest possible time and to live in it. So tenders were let for a quote from building companies. Although there are more than ten or so strawbale buildings in South Australia, I am told that mine is the first contract-built house in the state.

Building

The preparation of the site and pouring of the slab was much the same as for any house with the exception of the space for the atrium and a strip one strawbale width all around the outside which was the same level below the finished floor as the wet areas. Iron spikes about a metre high and about a metre apart were set in this strip.

Then a timber framework was built as the bales were to be infill. A platform of seven ply was built on the strip where the bales were to be placed; this was about 50 millimetres high and the spikes came through it. The seven-ply also went around the window and door spaces and it was all lined with black DPC (damp-proof course). Holes were drilled at about 500 millimetre intervals through the platform below the plywood and these were lined with half-inch garden-irrigation poly pipe to accommodate the high tensile fencing wire which was used to compact the walls.

WALLS

The bales were laid on the wooden platform — I was given the honour of placing the first one — and cut with an electric chainsaw where necessary to fit into the framework. At first they were impaled on the iron spikes. But as they rose above these, bamboo stakes were spiked into the straw to tie them altogether. This was because although the walls were constructed in the same manner as bricks they were not joined together with mortar. When the wall was tall enough — nine courses of bales — fencing wire was passed through the holes in the platform under the bales, over a topping then tightened down with wire-strainers and fastened with gripples.

To compact it, two hydraulic car jacks were placed on the topping board under a heavy piece of angle iron held down by a wire over each end. The jacks were then activated until the straw was compressed the required amount. The gripples were then tightened again and the two end wires shortened and tightened.

The next process was straightening the bales which was accomplished by a fair amount of what can only be described as bashing with sledge-hammers. The whiskers were trimmed off with brush-cutters resulting in absolutely straight walls. They were then covered with chicken-wire held in place by wire staples, about 200 millimetres long. To protect the straw from rain the walls were covered on the top and windward by plastic sheeting. A strong wind blew a lot of this down and it had to be redone one Sunday morning.

RENDERING

Next the walls were covered inside and out with cement render, two coats of cement mixed with lime, one thin and one thick, were applied on two consecutive days, then allowed to set and dry for three weeks. Because of the high lime content of the render it cracks as it dries and sets, and at least three weeks are needed before the final coat can be applied and this is expected not to crack. It didn't.

During this time, the remaining framework, the roof, all the windows and plasterboard were put in place. Then the final coat of render was applied. Once the interior was finished it was impossible to tell a plastered wall from a stone wall because the plaster was so straight and smooth.

All up, the coating of render was about 40 millimetres thick and easily withstood the 'kick' test by the engineer. The architect assured me that such a test would break the wall of a brick-veneer building. The walls easily withstood the 110 kmh winds that lashed the building just after the roof was put on.

DOORS AND WINDOWS

A big thrill was finding in a salvage depot the antique, stained glass front doorway from my former home in the suburbs. I was able to buy it back and I had it fitted into the house design. The house and doorway suit each other perfectly.

I have wide windows facing north, and the sun falling on tiled floors creates winter warmth. I ended up with wide windows facing all points of the compass so I'll need shade blinds on the east and the west sides of the house in summer.

I have a passion for light — and having lots of windows brings the scenery inside. Now, as soon as I open the front door, I can see down the hall, through the atrium, through the kitchen and livingroom, out through the bay window and down the valley beyond. It's just beautiful!

The bay window! It is an idea I bor-

The house at Mount Barker was built corner on, to take advantage of the views and to decrease the effect of wind on the house.

rowed from friends. They have a cast iron stove inside a bay window and one can sit in front of the fire and admire the view. I ended up buying a brand new, old-fashioned iron range for the same purpose and it has the advantage of a kettle constantly on the boil in cold weather and the oven bakes beautiful scones!

THE INTERIOR

The interior of the house was finished in a fairly conventional way, the floors are all tiled, two different sizes and shades of terracotta, with the exception of carpeting in two bedrooms. Flushing and plastering could be said to be standard. However window reveals are different: they are 300 millimetres deep extending into the room just beyond the skirting with no architraves around them as I did not want too much stained wood which would be too dark.

The ceilings are plain with the exception of very elegant ceiling roses and coved cornices in the sittingroom and hallway. In fact, the entire interior is painted a slightly off-white. Curtains, pictures and furniture provide colour. The white walls and the glass walls of the atrium mean that the house is very light because the sun streams into the kitchen and livingroom on fine days in winter no matter how chilly the wind outside.

But shade cloth is a necessity in summer, and I have hung a double layer of 50 per cent green shade cloth from the pyramid framework in the atrium as a temporary measure. All the potted plants are thriving.

The kitchen cupboards, the oven and the dishwasher are also white, but the bench and bar top as well as the bar exterior, all skirtings, architraves, window frames and linen cupboard are a light cedar shade. Other built-in cupboards are also white laminate which is so easy to clean. The bath, basins and toilets are white as are the wall tiles but I have a very colourful frieze around both bathrooms at waist height, and of course there are wonderful towels. Soon, I'll start on the garden.

Strawbale bears the load

The first loadbearing house in South Australia was not without bureaucratic difficulties. And yes, next time they'd do some things differently.

by Graham and Kaye Potter
Myponga, South Australia.

WE WANTED TO build a bed and breakfast cottage on our property which is situated between Myponga and Victor Harbour, and about 80 kilometres south of Adelaide.

The building had to fill two roles: first to be bed and breakfast accommodation; and second, we wanted to sell it as a house when we retired. It consists of a kitchen, diningroom, loungeroom, two bedrooms, a pantry and storeroom to service the bed and breakfast, a bathroom and a laundry. It has a return verandah, the back part being wide enough to park a car under.

The ABC's *7.30 Report* had a short program on John Glassford who was building with strawbales at Kangaroo Valley, south of Sydney. I signed up and went to a four-day workshop to see if this was what we were looking for. It was, so we got on with the planning.

LOADBEARING WALLS

We engaged Helen Bernard from Imagine Strawbale Constructions to develop the plans for the cottage. Unfortunately, it took nine months to get the plans approved by the council as this was the first strawbale house with loadbearing walls to be approved in South Australia.

It took eight months to complete, a month less than it took to get approval to build, from the time we started making the window and door bucks to the finished product, including landscaping.

Loadbearing design makes the wall building very simple as there are no posts to work around, but the disadvantage is that the walls need to be protected from the rain each night after working so we had to wrap the walls in plastic sheeting. The internal walls are all straw except the timber-framed one between the laundry and bathroom. This is where the water pipes were brought in from outside. It is not advisable to run any pipes through the straw walls in case of leaks.

RENDERS

Once the walls are rendered you would not know the house was made of straw, so we installed the obligatory 'truth window'. We also put in straw ceilings with exposed logs. These ceilings help to give the house a cosy feeling.

After installing the chicken-wire, the first coat of render was sprayed on and then the second and third coats were put on by hand. The surface was sponged off as the walls were not flat and couldn't be floated. The uneven walls and rounded corners help to give the house a softer look. Of course, when it comes to fixing cupboards and putting up cornices the uneven walls present problems, but it adds interest to the house.

Just to help you get things in proportion about the cost of strawbale construction, please consider these figures. The walls of the house amount to ten per cent of the total cost of the house. You won't save money by building walls of straw, but you will have a fantastically insulated house and a quiet house. You will be saving some trees by building with straw, but

Country Gates Bed and Breakfast has been very popular since it opened in 1999.

Below: The roof trusses and verandah hanger are put in place. The corners needed bracing to get the bales square.

there is still a lot of timber in a strawbale house.

We put a floating timber floor in the kitchen. Two weeks prior to opening we discovered that the floor really was floating due to a leaking pipe in the kitchen cupboards! Thankfully, the straw walls were started by first fixing treated pine to the slab, then putting the bales on top of this frame. This keeps the bales out of the water should there be any flooding during, or after, construction.

Finishing

We have found that the house seems to have a stable temperature. It is also a very quiet house. Would we build another strawbale house? Two-thirds of the way through construction we said, "Never again", but now that we've finished and can step back and look at the end result with pride, we have been heard to say, "Next one, we'll do it like ..."

If we built another loadbearing strawbale house we would put guides up on the inside walls like we did on the outside walls. We would install the top-plate on the internal bale walls at the same time as the top-plate on the external walls. This would help tie everything together and stabilise the internal walls. In theory, the entire roof weight is supported by the outside walls when using gang nail trusses and the internal walls are not to take any of the roof weight.

We would stipulate to the baling contractor that we don't want any rebales. Sometimes the twine breaks while producing bales and the contractor comes around the paddock again. He picks up the broken bail and re-presses it. This forces lumps of compressed straw into a new bale which tends to bend them, making building difficult.

Of course, we might go to a modified post and beam design. The main advantage with this system is that you can have the roof up before you bring the straw on site. This gives you, the builder, better working conditions.

You and the straw are out of the rain and the sun. But a post and beam structure requires more modifications to the bales so that they fit around the posts. Councils are more likely to approve a post and beam structure as they are familiar with this type of engineering.

We've gone the other way to the three little pigs story. Our first house in Adelaide was a full brick house, our second house was of pine logs and the third is strawbale.

House evolves from strawbale revolution

After about 60 weekends of work, a pole-framed shed turned into an appealing, comfortable family home that cost about $40,000.

by Des Menz
Clare, South Australia.

THIS IS THE story of how a humble building evolved into a house. Start with a shed plan that was to be the forerunner to the new family house, add some poles and a roof, construct some footings and build some strawbale walls. Then contemplate for a short time and think — this could become a house! Well, that's what happened, using recycled and plantation timber building materials, used and new fixtures and fittings, solar hot water heating, slow-combustion cooking and heating, wind and solar electricity generation. But I'm getting ahead of myself.

Leslie, our two children, and I left Victoria after ten years and moved back to South Australia. We wanted to find a block of land that would capture our imagination and challenge us in land rehabilitation.

We found a property only five kilometres from Clare, which is 130 kilometres north of Adelaide. The 33 hectares of hilly land had magnificent views but was suffering from vegetation clearance, over-grazing, erosion and the emergence of salt. None of these problems deterred us because we wanted to repair the land and build sustainable systems. My aim was also to demonstrate the integration of engineering and environmental solutions to land and water management, to construct a small grid-interactive wind farm, and one day to open the site for educational experiences.

SITE SELECTION

Our land is hilly and is very favourable for wind power generation. There is a major ridge at the western end of the property with a vertical rise of more than 70 metres from the lowest to highest point. When we looked for building sites for a shed and house, there were two north-facing slopes that gave good solar access but we chose the lower slope. This was for a number of reasons: better access, roadworks were cheaper, telephone cabling was less, and the slope was better for terraced gardening and for moving water between storages. So our site was 30 metres above our lowest point and about one-third of the way into the property.

Initially I had designed our house to lie in front of, and slightly downhill of, the shed (I have to call it a shed at this point), so excavations for both buildings were done at the same time. The shed site was benched into the rocky incline in a direction north-west/south-east — not the most favourable orientation for a house!

FROM SHED INTO HOUSE

The basic building structure is pole and rafter. I chose a mansard shape roof-line which is a conventional ridge roof with a large chunk of the top cut off, so that it

When the shed was a house: treated pine poles, which were the cheapest option, also satisfied building requirements.

Right: The bale wall, the slate floor and footings made of local stone.

mimics the hill shape and there is less wind uplift on the structure. Being an engineer I designed and built plywood box beams for the seven metre span and placed poles to resist wind forces, which can be quite severe.

Treated pine poles were used because of termite resistance, and because this was the least-cost option. As it happened, the shed became the house and pole-framing as the main structural element complied with the Building Code of Australia. A note to all would-be shed builders: always keep in the back of your mind that one day your shed may become a house, so compliance with the BCA is essential.

As the first few months passed — along with the first winter of construction — the basic frame of poles and plywood rafters were erected. Serendipity then descended on us. While we always wanted to try strawbale construction, we didn't choose this originally. But along came a new friend who just happened to have a couple of hundred bales for a cottage that didn't get built, and *The Straw Bale House* book.

After several nights reading the borrowed book we were convinced — let's use strawbales on the shed as a prototype for the house, and live in the 'shed-house' while we built the 'real' house. The borrowed book was returned, we bought our own copy, and we bought the bales.

The footings

There was an abundance of nicely cleaved stones lying on, and just below, the surface of our land which we collected. On the two sides near the excavated banks, we constructed formwork for stone footings to raise the bales well above the excavation surface to avoid possible damp problems.

With just a petrol generator for electricity, we mixed concrete in our mixer and placed this around the stones which were laid against the forms. Damp-proof additive was used in the concrete. The footings were made 450 millimetres high and 450 millimetres wide — the width of the bales — and were placed between the poles. The stone face has given an attractive feature both internally and externally. The footings on the other two sides were reinforced concrete 500 millimetres wide but only 150 millimetres deep, because of the very stable rock foundation. These were placed on the outside of the poles.

BALE TIE-DOWN

During construction of the footings, two different bale tie down techniques were used. The basic system was to pull a timber top-plate down on top of the bales to place them in compression, and thereby increase the wall stability. In the stone footings, ten millimetre diameter reinforcing bar was used and hooked starter bars were placed every one metre, ensuring that they avoided window and door positions.

With 900 millimetre long lengths of bar, I welded threaded rod to each end and made connectors by welding nuts to each end of 40 millimetre lengths of 15 millimetre water pipe. Although there was a little extra effort required, these connectors were much cheaper than those you buy.

For the concrete footings I wanted to try double-twitched fencing wire to pull the top plate down, so looped bars were placed in the footings near the footing edges.

WALLS

The walls on the stone footings were built in panels three metres long (the distance between poles) and three metres high (the height to the eaves) and were raised first to provide protection from the prevailing wind. The first course was placed on a plastic damp-proof membrane and over the starter bars. Subsequent courses were placed as bar segments and were connected together.

We drove 1.5 metre long bars (ten millimetres in diameter) through the bales at every third course and between the tie down bars. Bales were connected to poles with reclaimed flat steel fencing droppers; these were ideal because they could be bent to shape and they were pre-drilled for fixing to the poles. They were also cheap, and again slotted into our philosophy of using recycled materials.

In one section, I used these flat droppers in place of the bar tie down, bolting them together as the bale wall was raised. But I found it difficult to push the bale through the tie, so this system was discontinued. The problem with the bar system was the relative instability of the wall whilst working on it; having to take care in lining up the bales for neat fitting, and the difficulty in pushing the top course through the bar because of limited space.

The bales on the shallow footing were located on the outside of the poles to produce a continuous wall, and without doubt this was the quickest and easiest system, with comparable stability to the rod tie-down system. Again, the bales were vertically reinforced between each one metre tie down point, and also tied to the poles with bent flat steel fencing droppers.

The continuous wall required fewer half and part bales, so overall it would have been easier and quicker to have placed all footings on the outside of the poles. But there's always considered action we must think of! The advantage of placing the wall between the poles was the wider eave protection, and this is an important consideration for the bales.

WINDOWS

All our windows are recycled and were obtained from auctions and demolished local buildings. Two of our windows were bargains — cedar, worth many hundreds of dollars but bought for $74. Other windows came from a demolished timber building at the local primary school and were converted from spindle-hinged to side-hinged casement windows. The remaining windows were sold to us by local farmers, and were double-hung; after much effort on Leslie's part, they have been given a new life and have added to the charm of the building.

I made the window bucks from pine framing and 15 millimetre thick plywood cladding, to a width of 300 millimetre so

The south end of the house where loft space for the children was built when the shed became a house.

that when rendering the bales we could produce the sculpted rounded appearance around the windows.

The bucks were tied to the bale walls with the flat steel droppers, bent to shape so that they could be speared through the bales and screw fixed back to the buck frames. Where appropriate, windows have been connected to the poles along one edge. Overall, the windows are very securely fixed into the walls.

Top-plates

Timber top-plates were made to the width of the bales and to span from rafter to rafter, because the rafters sit on top of the poles. The function of the top-plates is:

- to seal the top of the bales;
- to provide a fixing edge for the chicken-wire sheathing;
- to provide uniform compression of the bales when rods and wire straps were tensioned.

The top-plates were made from a frame of 90 by 35 millimetre pine with cross pieces (stiffeners) every 500 millimetres. The threaded rods were anchored through the plate and a large timber washer that was supported by the stiffeners. Tightening the nut on the rod subsequently compressed the wall.

The wire straps were connected to threaded bars passing through the frame members, pre-tensioned by twitching, and again the nuts tightened to compress the bales.

I found that over several weeks the walls would 'accept' the compression load, and I would have to crank the walls down again. Finally, the top-plates were connected to the rafters to lock the whole wall in position.

Initially, I used the (you've guessed it!) flat steel droppers, but the better technique was galvanised steel purlin (off-cuts from the roof) reshaped into an angle cleat the width of the plate. These were screw fixed into the rafters.

Shed to house

Along with a lot of interior work and installation of outside services, and after about 50 weekends of building, we decided that the shed-cum-prototype house had had a great amount of effort put into it and that maybe it should become the house.

After all, we had raised the roof at the south end and installed an upstairs area for more space for the younger members of the family: this had taken much more time to construct than I had thought it would. Building the house had also become more distant, a fading memory, and we also wanted to reconstruct the old ruin of a cottage downhill at our entrance gate. We were also looking forward to just having a few weekends off! So the half-way house became the house.

But we had some misgivings. Firstly, the orientation was not ideal. While we have magnificent views, we have not achieved optimum solar access. But the orientation has some advantages: there is protection from the wind at our small front verandah; the shape of the roof allows the high speed wind to efficiently flow over it, and we achieve hours of morning sun in the winter. We are now planning to construct a bedroom and sunroom attached to the front side of the house, and this should improve the overall solar efficiency.

Secondly, it wasn't our ideal design. But with a little more attention to detail in the latter stages we have produced a comfortable building with efficient use of space. The lofty ceiling and mansard shape have added to the appeal of the interior. The thermal rating of the roof cavity — although only 400 millimetres thick — is R4.5.

Thirdly, we now have to design and construct other structures and the permaculture garden around the building. Rainwater tanks and a reed-bed filter were positioned for the overall house and shed arrangement, but we haven't found them ineffectively located.

Of the other building components, the composting toilet and shower rooms were positioned on the north-west corner of the house, and were clad in corrugated zincalume. Because we didn't have enough bales to finish the walls, the section over the main door on the east wall was clad in Ecoply over a double timber frame the width of the bale walls. This section was packed with insulation.

The gable ends at the raised roof section were also clad with corrugated zincalume. Insulation has been a high priority, as winters are cold and windy and summers are hot and dry. A slate floor and stone partition wall have been added to the interior for improved thermal control. Cross-flow ventilation in the house is very good and a ceiling fan moves air through the building during hotter days.

FINISHING

There have been about 60 weekends of work with very little outside labour, except for the very generous contributions over many weekends of a skilled friend, who I think just wanted to try out all the tools he had collected over the years! Towards the end of construction, some hired help was used to accelerate completion — we had to vacate our rented house — and there was some valuable help from friends.

The weekend work has included many other aspects such as installation of tanks; electricity system; pipework; drainage; reed-bed filter and all the things necessary for the house to function. Two-thirds of the floor area is stabilised crusher sand that was dry-mixed in our little mixer with ten per cent cement, laid out to dry, then watered in, followed by light compaction and finished off with a rich skim coat of sand-cement. The result was low cost and simplicity personified.

The remainder of the floor was covered in slate pieces from the nearby Mintaro quarry. This was Leslie's task and the end-product is very appealing and once again, low cost. Leslie says the grouting was a pain in the back! Our final tasks were to render the walls and get the solar-wind power system hooked up.

We have been enormously satisfied with the end result, and we have had a lot of fun picking up bargains at clearing sales and auctions.

The cost of the building, including the power system (bought secondhand but doubled in capacity with an additional wind generator and solar panels), tanks, drainage, toilet, solar hot water system, baker's oven (which is also our hot water booster in winter), but not allowing my labour cost, was less than $40,000. With strawbale construction, we have achieved a very comfortable and low-cost house, and one that has largely used low entrained energy resources in its construction. Just a lot of human toil!

VICTORIA

Been there, baled that

John and Rose have discovered some of the pitfalls, but above all the rewards, of strawbale building after doing it themselves.

by John Easterbrook and Rose White
Deans Marsh, Victoria.

MUCH HAS BEEN written about strawbale building. The thermal efficiency, the ease of construction and the environmental advantages are all well documented. The following are some observations that have come about from our own experiences and might give the bale builder some more insight into the process of building strawbale houses.

We bought our 1.6 hectare block in 1998 and we decided to build a strawbale house because we'd been talking about it for years. The proximity to the Otway forests and the extensive native agroforestry plantations in the area mean that bushfires are a real possibility. This was one of the main factors for our choice to build in straw. Studies conducted in America suggest that in most scenarios strawbale houses match, and exceed, conventional brick and timber homes when exposed to intensive fire.

Design

The Otway Ranges is an area that is usually classified as having a high rainfall but the last few years have been dry. The old adage "to give your building good boots and a good hat" is no truer than when one is building with straw, especially if using an earthen render.

If we were to do it all again we would not bother with an architect. After a four-month delay we drew and submitted the plans ourselves without a problem. Consider carefully who you will engage to draw your plans and know full well what you will require in your house now, and in the future.

We designed the house for passive solar benefit and with minimum walls — we knew we wanted space in the living and dining area, the kitchen and the study. We measured our bedroom in a house we were renting and decided that it was big enough. There is also a spare bedroom — with a substantial waiting list!

Some things were not planned to work out as they did. Our bedroom window faces east and that is the direction of the cooling breezes in the summer. From the bedroom we have a great view through double glass doors in the living area. The cross-ventilation will be more than adequate. We bought an external door for the bathroom simply because doors are cheaper than windows but it means that we can access the washing machine and the shower without bringing dirt into the house.

We have a 1500 millimetre wide verandah on three sides of our 15 square home. If you plan to render your house with earth a substantial overhang or verandah is imperative for the protection of your walls. The north side has an overhang which blocks the summer sun but allows the winter sun in.

Materials used

The internal walls will be hessian stuffed with straw and rendered with a gypsum-lime plaster. Our floors are poured, except for the bathroom, a mix of sifted road base with eight per cent cement. The floors, as well as providing thermal mass, give the house an old, established look.

The ceiling will be Solomit straw pan-

Rose and John decided to have the walls sprayed with an earthen render.

els, except for the bathroom which will be corrugated iron. The verandah posts, door frames and jambs, and gable ends are cypress which is naturally termite resistant.

We will use dam water for the toilet and we'll have solar power. We have a wood stove which will supply heat, cooking and hot water. It will be interesting to see how well it heats our home.

Some people complain that strawbale houses are too warm but if you incorporate energy efficient planning (small heating zones) with strawbales, then we're not surprised. We'll also have a gas stove and hot water service as back-up appliances. We've allowed for gas heating in the livingroom and study.

We decided to have a steel frame but unfortunately there were many hold-ups which meant extra costs and long delays. Steel frames also require the right tools and a different set of building procedures from timber. In retrospect, we would have been better off with a simple, timber frame erected by some local carpenters.

Walls

Once the frame was up and the roof was on we started laying bales, which was by far the easiest bit. The first course of bales sits on concrete footings 450 millimetres wide and 300 millimetres above grade. We used number four rebar pins inserted 300 millimetres into the footing and extending a further 300 millimetres above the footing to pin the first course of bales.

We did the longest stretch of wall in one day and we didn't get up early or work late, we made three trips to collect bales and we had visitors. The bale walls are tied to the footings with fencing wire and gripples secured to the footings with metal strapping (our west wall did twice fall down in strong winds).

Our biodynamic rice strawbales came from the Murray region. Their approximate dimensions are 450 millimetres by 900 millimetres by 400 millimetres. We both found the bales easy to work with, simple to cut and quick to erect. Other than the dust (a face mask is recommended) the bales were a joy to work with.

Doors and windows

All our windows and doors were bought secondhand, they fit well into the aesthetics of the house and were a real cost saving. It's well worth putting a lot of thought into window and door sizes and their distribution throughout the house.

All the windows and doors are flush with the external face with the following exceptions, the bathroom window and the kitchen window. It was necessary in the kitchen, otherwise we would have to climb over the sink to open the window.

An important consideration when choosing windows and doors is their height in relation to the top-plate. Strawbales are simple to modify lengthwise but the height is harder to modify. Therefore, it is a good idea to know the size of your bales so the windows and doors fit in neatly to the frame, minimising gaps and reducing the amount of loose straw and render needed to fill those gaps.

Rendering

Mice, mice, mice. Perhaps one of the strongest recommendations we can make to any strawbale builder is to consider how you will render your walls and do it as quickly as possible. The mice moved in after we erected half the walls and made a mess. Not only are your bales and your hard work at risk from rodent ruin, they are also more susceptible to damage from fire and rain.

The skin you choose to cover your strawbales with is vital to the proper maintenance and health of your house. 'Health' might seem a strange word to use but it became clear to us when we were building that we were erecting a living structure that needs to breathe and needs room to move.

For these reasons we chose to have the bales sprayed outside and in by Lachlan Burke of Quick-Straw Spray Rendering (see contact list on page 154). We were sick of mud during winter and spring but rendering, while hard work, was enjoyable. We were labourers for Lachlan to save money. The first two days of rendering were hot — 39°C — but we survived. This was money well spent as not only have we saved an enormous amount of time but we also achieved a finish that allows the walls to breathe as well as being a joy to behold.

We used a combination of clay and lime applied in successive coats with the final coat being brushed on with a soft broom. One of the greatest advantages of the clay-lime mix is the ease with which any faults can be repaired. There is a lot of information available about rendering strawbale buildings and each method has its pros and cons. Regardless of the method you use, strive to render your bales as quickly as possible.

What we've learnt

We are currently at lock-up with three more rooms to have the floors poured, internal walls to be completed and fitting out to be done. So far the project has cost us $40,000. We believe that strawbale building is no cheaper than a contemporary brick or timber house other than the savings in labour if you choose to build your house yourselves. Walls only comprise ten to 20 per cent of the total cost.

The true cost-saving with strawbales is that bale building allows anyone to have a go. You do not have to be a specialist and, in this day of specialists, it's nice to know that there is a means by which everyone can lend a hand in building their own or someone else's house. The other significant saving is in the reduction in heating and cooling costs.

One thing we neglected to budget for was the cost of earthworks. We hadn't really thought further than the footings. It pays to know beforehand exactly what needs to be done, for example, the septic tank, distribution trenches and drainage trenches.

It is imperative when dealing with any person whether they are the plumber or the drafter that you both have a clear idea about the nature of the work, the expectations of the finished project and, above all else, the final cost.

If you become a strawbale builder or a strawbale home owner, you will be in essence a pioneer in a relatively new building technique. Some tradesmen can be very conservative when it comes to anything they consider to be outside the norm. Therefore take the time to talk.

Perhaps the single best thing about building your own bale house is that it looks and feels great. Our eyes never tire of watching the shadows shift along the contours of the house. The flowing form of the house is very relaxing and a great break from straight, sterile, conventional plasterboard walls. Children especially seem intrigued and delighted by the deep window sills and curves of our bale house.

Being an owner-builder, while frustrating, has its rewarding moments. If we had contracted someone to build our house we would move into an environment that was strange to us. But we already know our house and spend quite a lot of time in it, working and relaxing. The first time we saw the moon shining through our windows and doors was exciting. Every day we monitor the movement of the sun across the floor. The most important piece of advice we can offer is, after you've read your chosen material, try it for yourself.

Innovative materials need imaginative design

Helen and Per have an architectural and building practice that focuses on energy efficient buildings.

by Helen and Per Bernard
Imagine Strawbale Constructions
Daylesford, Victoria.

WE HAVE AN architecture and building practice with a focus on energy efficient building. When we first saw the beautiful images in *The Straw Bale House* book we were immediately attracted to the soft but strong forms created with strawbale construction. Per travelled to America and worked with builders who had completed many strawbale buildings. A community of enthusiasts welcomed him, and generously shared their experiences. Imagine Strawbale Constructions was established in response to the growing interest many Australians have in this building method.

We live in central Victoria, where the winters are cold and frosty, and even in the summer, sunset is followed by a chill. Having lived in old timber cottages characteristic of the area, built without consideration of the sun's ability to warm a home, we had first hand experience of the high cost of heating these buildings, and the poor level of comfort they offer. Having built our own strawbale home, we now have the pleasure of living in a strawbale home.

Our strawbale home offers high levels of insulation in the walls, matched by an equally thick layer in the roof. Very little heating is required to keep this family home warm, the thick walls are cosy and comforting. Inside the home is a sense of calm and stillness, often broken by the sounds of children playing. A house designed in response to the environment is wonderful to live in. We have designed, and sometimes built, many strawbale buildings. These include the following:

Loadbearing method

The simplicity of Country Gates Cottage lent itself to the loadbearing method, although it could have been done by the post and beam method. It is one of the few loadbearing houses built in Australia to date. The engineering was straightforward and certification of the wall system was provided by the engineer. This was not good enough for the building surveyor however, and obtaining a building permit took months. This was due to the building surveyor lacking familiarity with strawbale construction and the laws of South Australia, in which the cottage has been built. The cottage is now operating as bed and breakfast accommodation, offering two bedrooms and open plan living (see page 80).

Modified post and beam method

Stony Creek House is a compact, two storey house, oriented to the northern sun. Pergolas on three sides will be planted with deciduous, food-bearing vines and fragrant climbing roses, providing shade in the hot months of summer.

The strawbale detailing incorporates soft curves around doors and windows internally. Window sills are finished with timber veneer, incorporating window seats large enough to sit in, most having lift up lids for storage. A truth window with a rendered frame is there to remind one of how these embracing walls have been created. The render is cement with Porter's limewash. The upstairs flooring is limed plywood. Per has carved an arched niche in the stairwell, using an angle grinder with cutting blade. A soft light is set into the base to guide the way up to bed at night.

The house is heated with a slow combustion stove, which also heats the hot water (solar hot water in summer). The stove is more than adequate to heat the home comfortably. Because of the high thermal mass which is provided by the concrete slab, internal walls of Mount Gambier limestone, and the thick layer of render on the inside of the strawbale walls, it is possible to air the house, to sleep with windows open to the forest air and stay warm inside.

Double glazed timber windows and doors have been fitted with seals, to prevent air leaking in and out around them. The amount of fresh air coming into the building is controlled by leaving the window slightly open.

Hilltop Cottage is a two bedroom accommodation house nestled into the side of a hill, overlooking Lake Daylesford. Because of the steep site the building has been set on stumps, demonstrating the remarkable flexibility of strawbale construction. Lofty ceilings and high windows flood the interior with light. Cosy corner window seats in the bedrooms provide a sunny place to settle into with a good book.

The Mediterranean House is the first professionally-built strawbale house in Victoria. It has comfortable indoor temperatures through the scorching days of summer and winter snows. The frame of this two storey modified post and beam house has been built with pine. Strawbale was the ideal way of evoking the thick solid walls of Mediterranean houses in an energy efficient way.

The home has been awarded a five-star energy rating from Energy Efficiency Victoria, scoring double the number of points required.

The Strawbale Gallery is the strawbale exception which proves the rule. Instead of placing the windows to maximise solar gain, the Strawbale Gallery uses fenestration to provide natural daylight to showcase art works in the best light. Stable indoor air temperatures offered by strawbale buildings are ideally suited to housing art works.

The ability of strawbale walls to maintain constant indoor air temperatures has other commercial applications. The team at Imagine have designed wineries, fish farms and restaurants which are viable because of their high insulation value.

PRINCIPLES OF PASSIVE SOLAR DESIGN

Our strawbale homes have been built with these general principles in mind.

Orientation. Most of the glass is on the north, where winter sun can warm the house and summer sun can easily be shaded. Where views are available in other directions we depart from the rule, and bring in other elements to compensate, such as a pergola with a deciduous vine or an external awning.

Insulation. The thick, super insulated walls of a strawbale building have at least four times the insulation value of conventional constructions. Even the gap between the window or door frame and the wall it is inserted in, is filled with insulation and caulked to provide improved levels of insulation. Just as the walls are super insulated, so are the ceilings of our buildings with a standard 450 millimetres of insulation.

Thermal mass. When massive elements such as earth, brick, cement or stone are provided inside the insulated building, the temperature inside the building is stabilised. Warmth from the sun, or another

Stony Creek House is a strawbale house that won a five-star rating from Energy Efficiency Victoria.

Right: Building Hilltop Cottage: the chicken-wire is in place, the windows detailed and bale stacking begins.

heat source, is stored in these elements and released into the room when the air temperature drops at night or when the heating is switched off. Strawbale walls have a good thick layer of render on the inside of the walls which serves this purpose beautifully. A concrete slab, and some massive internal walls add to the beneficial effect.

Infiltration. Air leakage around windows and doors, and at other joints in conventional construction adds considerably to the amount of energy often pumped into heating and cooling these buildings. Strawbale construction can reduce this, as a layer of render extends from below the slab edge up to the ceiling or eave without a break. Windows and doors are fitted with seals, so that the amount of air that leaks in or out of the building can be controlled by the occupant.

Cross Ventilation. Windows, when placed on opposite sides of a room, and of the house, can be opened to provide effective ventilation of the interior. This allows cooling in summer, and freshening of indoor air all year round to maintain a healthy indoor environment.

Four months and it's finished

It's vital to do research before you start, says this owner-builder, who built a comfortable family home in the suburbs.

by Steffan Klein
Mornington Peninsula, Victoria.

THE FIRST TIME I read about strawbale building was in the early 1990s, in *The Age* newspaper. Patrick, our second son, was on the way and the house was too small. Wow, I thought, here is a building method that makes sense. Easy to do, cheap, and environmentally friendly. At least that's what the story claimed.

However, when getting an approval turned out to be far too complex, we ditched the plan and built a normal house. Of course, I couldn't completely forget about building with straw. The promises of easy building and low cost were simply too tempting. So in 1997 when the building bug bit me again, it was going to be straw. My wife and I had decided we wanted a comfortable house, a place you would walk into and immediately feel at home. Straw seemed to be the right stuff for all these reasons.

By now I had ordered instructional videos from the United States, some 'how to' booklets and I had done quite a bit of research on the Internet. What struck me about the Internet was the amount of people who had obviously never built a shed, let alone a house with straw, enthusiastically dishing out a lot of information. It was usually well meant but often of little help and even misleading.

I needed to do more research in order to distinguish between realities and fiction. In the beginning all the information I came across argued that loadbearing strawbale construction — in which the bales support the weight of the roof — was the best way to go. Loadbearing strawbale walls were more environmentally friendly, they saved timber, they were easier to build, and a real strawbaler would never build a house on posts were some of the claims I discovered.

Reality check

Then came *The Straw Bale House* book, and with it one of the most important lessons: enthusiasm is great but a reality check is better. Loadbearing strawbale construction can use more timber than the modified post and beam method. After all, I was going to build on Victoria's Mornington Peninsula which virtually guaranteed rain during the building period.

The decision to go with the modified post and beam system solved the problem of keeping the bales dry. We would simply get the bales delivered after the roof was up. Never was I gladder I had made this choice than when the bales actually arrived. Until this moment I hadn't really thought about how much storage space the bales would take up before they went into the walls.

The first shock came when a huge semi-trailer, loaded six bales high, turned into our street and stopped in front of the

The Kleins' house evoked much interest in the suburb.

Right: Steffan found spraying the render saved a lot of time and strongly advises against rendering by hand.

house. Out came the calculator because obviously I had ordered far too many bales. I hadn't. The second semi-trailer arrived 15 minutes later. Fortunately the trucks could pull up right next to the house. The great skill of the drivers, who virtually threw the bales into the house, made stacking the straw far easier than we had feared. You will not have this advantage without good access.

Disheartening

The house was already quite advanced by the time the bales arrived. The slab was done, the roof was up, the initial plumbing and other connections were provided. At this point we had already sunk around $40,000 into this project — and our future home still only looked like a barn — especially with the straw in it. This was probably the most frightening and disheartening time during the building process. We wondered if we had made the right decision.

We had started with a limited time line and a set budget. Much of this money was already gone, yet there was little to show for it. Professionals had done the slab and plumbing. A labourer and I put in the tubes for the hydronic floor heating — which is great, by the way, and better than any other heating system I investigated. A carpenter and his helper put up most of the frame with the labourer and me helping. Once the posts were up, and with them the roof beams, the labourer and I put up the roof, and then waited for the bales. To stay within the time frame of under four months, it could not have been done differently — yet a real house still seemed a long way away.

I could probably have saved more money by doing more myself, but this would have taken longer and I would have probably spent the savings on renting a house. Some tasks I left to the professionals due to a lack of time and insufficient experience on my part. I was not going to risk the quality of our house.

Raising the walls

By now the big moment had arrived. The straw was there, the neighbours were wondering, and a group of volunteers from a local horticultural group came to help us raise the walls one weekend. Despite some disorganisation and a steep learning curve, we managed to raise the walls for the front third of the house. For the rest, there were only two of us. But by now I had overcome my biggest worry — how difficult would it be to create halfbales (very easy actually) — and so we just forged on.

What was much more work and much more time-consuming than I anticipated was wrapping the walls in chicken-wire to get more support for the concrete. To save time with the rendering, I had organised a two-man team with a concrete pump to help us render the walls. After initial start up problems this worked extremely well and was very fast. When we tried to render an office building a year later by hand we gave up in frustration after a day and used the spray gun again.

Advice on rendering

If you ever read those stories about the joyous time everyone had rendering the strawbale wall raised during a workshop, try to imagine how joyous it would be to do the same wall three times, and not just a small shed wall but a complete house, inside and outside. And to actually finish it and not leave when the volunteers had enough. It takes a long time to render by hand and it's extremely hard work, especially if you are not used to it. If you can afford it, get someone to pump the render mix onto the wall and just organise some help to smooth the walls. A so-called sponge finish is easy to achieve once the render is on the wall. We had never wanted straight walls so we didn't have to worry about the uneven surfaces.

Finally the place started to look like a house. Now, two years later, the house is finished, the paint is on the walls, the plants have grown. When we moved in after the four-month construction period, the inside of the house was only missing doors (apart from the house and toilet doors). This changed when my father-in-law visited. He put in all the other doors and finished off some things that I hadn't had time to complete. Some months later my father helped paint the outside. Actually he did the lot while I went on a holiday with my wife, and my mother looked after the kids. Wow, this is how work should always be done!

Dream home

By then, the house had turned very much into our dream home. The house feels comfortable to us and to many of our visitors, the minute you walk in. It is about 18 squares, so it's not a small house, but for a family of five it has a nice size. I am now sitting in my strawbale office in the backyard, which we raised about a year after finishing the house.

Here I put to use some of the things I had learned while I built the house. For example, we did not bother hammering metal rods into the bales to connect them — there was simply no need for this because the chicken-wire and concrete hold the bales in place. Instead of flat we put the bales on the side, achieving the same R-value due to the way the straw is packed, but wasting less floor space and needing fewer bales per square metre, which made building even faster and helped cut costs. And we followed the modified beam construction method even more closely, which made everything easier.

Would we build again with straw? Yes, definitely. This is not just me saying this but my wife. The house feels great and

The swimming pool, the palm and the classic lines of a strawbale house.

the insulation of the straw works like a charm. In two winters the heating never ran for more than two hours a day, if at all, and the temperature never fell below 20°C during the day. Our gas bill is down by over 60 per cent, despite using gas for cooking, heating and hot water. So even though it wasn't cheaper to build the house, it is definitely much cheaper to run it.

The only problem: I wish I had insulated the roof even better. In summer when there are many hot days, it heats up and eventually the heat penetrates the house. Once inside, the good insulation can actually work against you. Another issue: it is hard to keep the place cool, when even cooking can create enough heat to warm it up. Roof windows, which we can open, have solved the problem of excessive heat generated or trapped inside the house.

TIPS FOR BUILDERS

All in all building with straw was fast and relatively easy. I am glad I did not listen to all those over-enthusiastic people and did my own research — and I would recommend that everyone does the same. I would always recommend using the modified post and beam construction method. Space the posts every 2.5 to three metres — that will stop the bales from toppling over until you have the chicken-wire in place. Using this method will also make it easier to get building approval from the council, another big advantage.

Some of the biggest worries about building with strawbales can be allayed simply by getting a few bales before building and fooling around with them. Just build your own straw needle and create a few halfbales by following instructions in *The Straw Bale House* book. You'll soon see why so many mudbrick owner-builders were absolutely fascinated when they saw how quickly our place went up.

Do we have any regrets about building with straw? None whatsoever. We built in the middle of an established suburb and had a huge amount of people approach us because they were utterly fascinated by what we were doing. Well, I do have one regret: I should never have listened to all those negative comments and done it much earlier.

We did it our way

Sue and Don built a workshop/studio on their land near Daylesford using initiative and imagination.

by Sue Ewart and Don O'Connor
Daylesford, Victoria.

WE ARE owner-builders with limited financial resources. But this is the only resource in which we are limited: we have imagination, energy, friends, initiative and a spiritual awareness. All of these have enabled us to create a building that serves on many levels without having to join the 'real world' of high finance and 'nine to five' employment. Our limitations of finish, finesses and fastidiousness are just that — our limitations — and thus we accept the way a particular door is hung, a window is set or a wall surface is finished. We accept the responsibility for our building.

PLANNING

Our project involved the construction of a large clear span area, an office area and an upstairs studio. We also added another room that will serve as a shop and display area. The total area is about 27 squares (around 250 square metres). So far, it has cost us less than $1000 a square!

Building is a process of research and decision-making, followed by resourcefulness and sensible spending. You must be able to dream the dream. Put down on paper what you would really like and then work out how. Not being trained builders we are able to have a go in very different ways to the norm and we believe this has worked for us. Often when you are trained in an area you cannot vary outside of the formal rules.

We sketched a shape and a roofline, Don always wanted a 9 metre (30 foot) width and a 12 metre (40 foot) length for the workshop space so the other areas fell into position around that. Costing then came into it as we researched what was available. We almost gave up on the roofline because experts quoted us $15,000 or more. It eventually came in at $7,500 and this included the frame.

To achieve the desired span the amount of wood necessary was frightening, a veritable forest of trusses and timbers not to mention the woodworking and framing skill required. Thus we looked at steel — in hindsight it would be secondhand steel but at the time none of the 'experts' hinted that we could use secondhand steel and still get permits.

BUILDING

We started with a steel portal frame spanning nine metres. Four of these spaced six metres apart linked by a footing of 550 millimetres by 500 millimetres with the appropriate rebar. The two frames at one end had post heights of five metres, the rest being 3.1 metres. The height was to accommodate our upstairs studio. This design appears very spartan but with the centre bay cross-braced with 14 millimetre rod it was quite stable and met the engineering standards. Steel C-section purlins joined the frames forming a base for the roof of zincalume. A big hay shed, in other words.

Into the wet concrete strip footings we embedded threaded 12 millimetre rods, with 100 millimetres bent at right angles in the concrete and beneath the top layer of trench mesh. We also inserted old rebar at intervals between these threaded rods

Doors and windows were gifts from friends and family and salvaged from demolition yards.

Right: Don positions rebar into the footings.

to assist in the location of the first and second row of bales. The threaded rod was spaced about 1.5 metres apart, allowing for doors and windows.

The technique we used was called compression even though we were actually infilling the structure. We compressed the bales in the wall between a top-plate and the foundations. We chose this technique because the distance we had to span between the uprights was so great. A post spacing of three to four metres would have enabled us to use plain infill without the compression. The large spaces we had between posts was a major dollar saver in our building. Big sections of wall being built quickly and easily without the need to have other framing.

The method of compression we chose was to use threaded rods, embedded in the footings, to protrude through a wooden top-plate.

This is a reversible process so should we have a change of mind about walls or windows we could accomplish this without any problems or wastage. A large washer, a nut and a spanner completed the setup. By screwing down the nut the compression plate was used to apply pressure to the bales in the wall, turning them into a solid, compact mass. Over a height of six bales we expect a compression of about 10 centimetres. We saved lots of money by using heavy old timbers from the demolition yard for the compression plates.

Damp-course, bales, top-plate — that was our wall sandwich. We used brackets and tek screws to attach the top-plate to the steel posts. This creates a solid, compact straw wall that would defy all but a bulldozer to knock over. We used this form of wall construction on all of the outside walls, including the upstairs section. We used a straight infill technique (no compression) for one of the internal walls and also for the amenities block, another steel-framed building.

We inserted wooden stakes in the infill

walls to locate and strengthen the bond between the bales.

RENDERING

The earth from the building site has a very high yellow clay content and this proved ideal for the first coat. As we added the second and third coats we increased the amount of sand in the plaster mix to reduce the shrinkage and cracking. By the time the third coat had dried we were left with a hard but gentle surface that wears well, resists rain and looks great, retaining all the colour of the original clay. This type of finish is practical, inexpensive, time consuming, unskilled and lends itself to a part-time approach.

Earthen rendering is a very organic process and you have to play around with your recipes. We started with recipes given to us by others with experience and have changed them to suit our clay, our weather and building conditions. Dry, powdered clay was easy to sieve but a pain to mix with water if we needed it that day, however, soaking it for a few days gave us a nice, sticky base mixture.

Our area is very wet most times so we worked out a slurry and sieving system for wet clay. At first, we finely sieved the first coat and then sprayed it on the walls, but we have since discovered that we can just rub this gooey clay mixture in by hand. Chopped straw and cow manure are two other additives of the wall render mix. The straw helps bind and strengthen the render similarly to how it works in cob, and the cow manure is another waterproofer and binder. We prefer using the off-cut straw from trimming around doors and windows, or chopped straw made by putting straw in a big plastic barrel and poking the whipper snipper in and giving it a good chop.

We used this earthen render finish on the exterior and interior walls. We put flour-paste glue on the inside walls to minimise the amount of dust, and it gives a low sheen to the walls. We put vegetable oil in the mix for the outside walls which increases resistance to the weather. We were further encouraged with our earthen rendering when we met up with Keith from Rico in Colorado who has followed a similar discovery path and his recipes are almost the same.

On many of our workshops we have had engineers desperate to mechanise the process. The concrete mixer does not work for the render as the clay gets stuck around the blades. We have however a brilliant appliance (homemade by an engineer on one of our workshops) that is a piece of steel purlin bolted onto a stem of threaded rod looking similar to a paint whipper. This attachment fits into our electric drill. This gadget can be poked into tubs to assist in the making of a good base slurry, the sharpness of the steel cuts into any clay lumps we have and has saved lots of time. When the blades wear out a new one can be made by cutting off more purlin bits using the angle grinder.

Windows and doors were fitted later into rough, wood frames that are placed into the walls as they go up. Very basic, very simple.

Most of our windows are gifts from friends which is a bit like a patchwork quilt with the loving energy of all your friends, a few of the larger ones are from our favourite demolition yard and the owner is now also a friend.

Purchasing is a very important part of the project and it should be tackled seriously but with a sense of fun. Compare prices and be willing to barter with business people, treat business proprietors as friends, chat to them, learn from them and create a relationship with them.

We have developed great trading arrangements and friendships with the farmer who supplies the straw, the demolition yard owners, the local cement business, the roofing supplier, the local hardware merchant to name a few. Be willing to tell them your plans and ask for advice: you don't necessarily need to take it, but it gives you more to choose from. Ask for owner-builder discounts — they can only say no — and enjoy the whole experience.

Our strawbale philosophy

Strawbale should be regarded as a liberating, empowering building material and treated differently from conventional materials.

by Sue Ewart and Don O'Connor
Daylesford, Victoria.

AS IMPORTANT to us as the physical process of strawbale building so is the importance of keeping the building and its process in perspective. We are building a structure to protect us, to provide us with a secure place where we can work in comfort, away from the weather, safe in our house. Do we want to look around at a scene that does not relax us? The following are a series of questions we feel everyone should consider:

- Do you want to live in close proximity with a beautiful surface that off-gasses toxic chemicals?
- Do you want your building to support you, heal you after your exposure to the workplace, the city, and pollution?
- Do you want to live in a stereotype house?
- Do you want to live in a house that reeks of the energy of a disgruntled worker, stressed by deadlines and the need to cut costs at every turn?

Houses have another potential to fulfil. They are not necessarily the place where we go when we leave work, the place where our bed happens to be or where we keep our Wheaties. Instead they can imbue us with spirituality. They can support us in our beliefs, bless us, and teach us.

In the native traditions that we have studied, a house is put together in a prayerful and considered manner. Symbols are incorporated into the buildings to send messages to the creator and ancestors and the building is in harmony with its surroundings. We believe that has been achieved in our strawbale building and can be achieved in most other buildings, benefiting all who inhabit them.

Advantages of strawbale construction for the owners:

- Freedom of design;
- Softness of structure;
- Sculptable;
- Potential cost savings.

Advantages of strawbale construction for the builder:

- Ease of construction;
- Fast training of workers.

Advantages of strawbale construction for the planet:

- The construction materials are benign, natural and plentiful products instead of energy intensive, toxic and non-renewable items;
- Far less destruction of planetary systems to produce the materials needed for building;
- Less dependence on artificial heating systems;
- Less stressed human beings;
- Aware people;
- More shelter at less cost to the planet.

Sue and Don's building started off with a shape, then a sketch. The advantage of owner-building, they say, is that you are free to try new things, new methods and new designs.

DESIGN

Strawbale is unique in its properties and should be treated as such when designing. If we ignore this and design as we traditionally have done we will end up causing more problems. Strawbale is a material that lends itself to use by the people.

Technical over-indulgence is detrimental to the widespread use of strawbale as a positive building technique.

We are both keen to build a house that is a mixture of building styles and materials. Strawbale construction lends itself to be used alongside cob, mudbrick, and rammed earth. An excessive amount of wood or steel framing ignores, and abuses, the potential of strawbale construction. Many buildings are being built with almost traditional framing structures and the straw is an infill which is more like an insulation batt. The walls are then wrapped in wire and cement rendered. A wood or steel frame cement wall with some straw in between.

A lot of people are clamouring to be the first straw this or the first straw that. Our society needs to get off this — straw has been a building material in Australia for over one hundred years, although that has not been widely publicised. In our travels and at field days we have met many wonderful and informative old timers with wonderful stories of helping their grandparents build and render their straw house.

ACCEPTANCE

Will people accept the functionality, comfort and beauty of buildings before the questions of cost and status? Will the building industry push for more restrictions or will people be encouraged to stretch the boundaries of design with this new but old medium?

This will ultimately be determined by how much people are willing to be dictated to by others.

Winery builds country's biggest

Strawbale appears the ideal material for providing a stable environment which is essential for storing quality wines.

by Bridget Goodwin
Canberra, Australian Capital Territory.

VICTORIAN vignerons Monica Vineyards have constructed a massive 250-square metre winery and tasting room out of jumbo strawbales in what's believed to be the biggest structure of its type in Australia.

The building at Lethbridge, near Geelong, has a concrete slab floor and walls made up of 220 jumbo strawbales creating 4.6 metre walls. The structure is loadbearing — the strength of the bales themselves hold up the roof. Steel windows and doors frames have been designed by an engineer to fit into the building design which will be spray rendered with a tinted concrete finish over strained wire and netting, anchoring the bales to the slab.

Monica Vineyard owners Ray Nadeson and Maree Collis, with their baby.

Susan trims the walls with a whipper snipper.

Monica Vineyard is managed by pharmacologist and physiologist Dr Ray Nadeson and his wife, Dr Maree Collis, a synthetic chemist.

They began planting their Lethbridge vineyard over three years ago and have been searching for a suitable construction to process their grapes, store barrels and hold tastings for cellar door visitors. Researching buildings on the Internet, Ray came across information about strawbale building and contacted the Australian company Huff 'n' Puff Constructions.

"It's my belief that good wine needs to be stored in a stable environment so I looked for a building system which can produce wine of quality," Ray said. "If you look at the French they've been able to store their barrels in caves and I thought we need to do the equivalent. This is an above-ground version.

"I must admit that at the time I thought, 'how the hell am I going to do this?' but I contacted John Glassford and Susan Wingate-Pearse and they came to visit us and inspired us to try this sort of thing," said Ray.

The jumbo 2.4 by 0.9 by 0.9 metre strawbales are super insulated and are capable of maintaining a constant temperature of 17°C. The walls of the winery were put up in three days using a front-end loader with a team of helpers straightening up the walls, five bales high, and helping to position the steel door and window frames.

According to John Glassford, strawbales are a logical way to construct wineries and commercial buildings of this type while serving sound environmental concerns with high insulation (which reduces heating and cooling costs) and their capacity to lock up carbon from the atmosphere.

"It's the most logical thing. It keeps running costs low, the insulation factor is huge at R20, and we've locked up 35 tonnes of carbon," said John.

Bed and breakfast in a cold climate

Strawbale has fantastic insulating qualities as Helen and Per have discovered.

by Helen and Per Bernard
Imagine Strawbale, Constructions
Daylesford, Victoria.

WORLDWIDE interest in strawbale house building is growing. Strawbale buildings exist on every continent, both as low-cost housing and more mainstream buildings. Down Under, activity is really gathering momentum. During winter 1998 we designed and constructed our first strawbale building and now, being more enthusiastic and a little wiser, we've built and/or designed quite a few more. Some of the latest include a gallery and a winery.

The Last Straw

We have been encouraged by the international strawbale community, and inspired by its journal, *The Last Straw*. Here we are reassured, taught and generally informed by those who have tried, tested and experienced strawbale construction. Through our work with experienced strawbalers in the USA and visiting some of the thousands of completed houses there, we have gained the confidence to offer strawbale construction in our architecture and building practice.

Bed and Breakfast Home

Our first strawbale job was to design a house which also offers bed and breakfast accommodation, at Daylesford in central Victoria. In our initial discussions with our client, we mentioned our new interest in strawbale construction. A brief description of the thick walls, deep window sills and exceptional insulation qualities were enough to inspire her to go for strawbale. And so our journey began.

We met with a structural engineer, Peter Yttrup, who is interested in alternative building technology. He has designed several strawbale buildings in recent years. We decided to build the house using the modified post and beam method. This avoided the complications of convincing the authorities of the potential of loadbearing strawbale construction. After several discussions we had the full support from Mount Alexander Shire in issuing the building permit, as well as their positive interest throughout the construction. They are keen to promote strawbale construction.

With building permits, insurances and contracts in place we were eager to start, within six months of receiving the commission. We poured a conventional concrete slab, closely followed by the wall framing. We used plywood box beams, the thickness of the bales, around door and window openings, while ten centimetre posts were located at each corner to support a roof pitching beam. We used long span, posi-strut joists for the first floor framing, and conventional gang nail trusses formed the roof. As a 24 square, two storey home, the framing was a big job.

Getting the Strawbales

Once the roof was on, we had somewhere dry to store the bales and half the bales

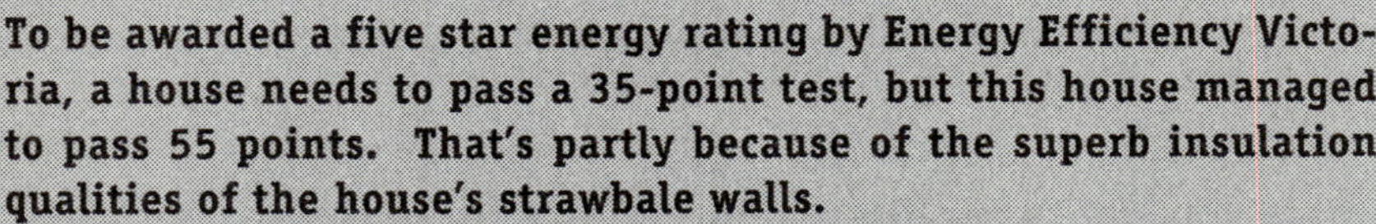

STRAWBALE HOUSES

Heating and cooling costs

To be awarded a five star energy rating by Energy Efficiency Victoria, a house needs to pass a 35-point test, but this house managed to pass 55 points. That's partly because of the superb insulation qualities of the house's strawbale walls.

An insulated weatherboard house has an 'R' rating (resistance to heat flow — the higher the 'R' rating the better) of about R2; a brick veneer house with no insulation is about R1.5; a double brick house with foam cavity insulation is about R1.5; and a mudbrick wall is about R0.5.

The Daylesford house's strawbale walls are a staggering R8. Per Bernard points out that 65 per cent of a house's heat loss is through the ceiling so there's no point having strawbale walls with poor ceiling insulation. The house's ceiling insulation — recycled pulped paper (cellulose) — is rated at R6.

So what does this mean for annual savings on heating and cooling costs? A five star energy rating means the 'average family' will save a massive $1200 per year on heating and cooling costs, or $20,000 over the 'typical' mortgage of an average-sized home.

were delivered. Regular bales of wheat straw were supplied from a baling contractor in Elmore, Victoria. A local farmer lent us a conveyor to carry the bales to the upper storey. We had installed chicken-wire on the outside of the frame, stretched taut so that the bales could be stacked up against this plane. No bale pinning was required because the straw was simply infilled within the post and beam frame.

After the bales were in place out came the whipper snipper, which easily trimmed the bales to a neat finish. We then fixed chicken-wire to the inside and used a bale needle to stitch the two sides together. The chicken-wire held the bales in place, despite strong winds. We installed expanded metal or diamond lath, finished with plasterers' stopping bead, around openings and to cover any exposed timber that was to be rendered over. Then we started the rendering.

RENDERING

The house is hand-rendered inside and out with two coats of cement render. When applied over chicken-wire inside and out, the cement becomes reinforced cement skins, to form a sandwich panel of great strength which is insulated with straw. The cement is water-proof but breathes to allow condensation vapour to escape from the strawbale wall. The option of earthen renders did not appeal to the owner who wanted a low-maintenance building. Because the house is in a high rainfall area, the benefits of a cement finish were more appealing than climbing a ladder regularly to re-render two storey walls.

On the outside, we finished the walls with a cement colour coat. The product we chose was free from bonding agents and sealants, to maximise the ability of the wall to breathe.

Internally we finished the walls with

It snows in Daylesford but the house is warm inside.
Right: Trimming the bale walls with a whipper-snipper.

serene lime plaster and beautifully coloured limewash. The texture of the walls varies from room to room because we trialled different techniques. The owner enthusiastically and carefully detailed some of the render in the deep reveals herself, with her own personal touches such as a basket weave pattern.

One afternoon a curved and stepped landscape wall was built and rendered another day. This time we used a spray render which gave a quick, strong and nicely-textured finish that complements the smoother finish of the house.

High ratings

Energy Efficiency Victoria have awarded the house a five star energy rating, the highest possible. The passive solar design incorporates north-facing windows shaded by pergolas in summer, highly insulated walls and ceilings rated at least R6, as well as thermal mass provided by the tiled slab and the thick render on the walls. The hot water service is in the laundry where its heat dries the washing and the room. Two small wood heaters warm the house in winter months. The owner does, however, have a problem with these. You see she hasn't been able to find a way to get the heaters to burn at a suitable level without getting too warm!

To be inside the completed house is a wonderful experience. Thick, solid walls have a calming effect. They are slightly uneven, and gently rounded at window openings and corners. It is very quiet. The rendered finish gives a timeless feel to the house.

Compared to other building methods, strawbale construction is excellent value for money. It costs no more, but has wonderful, unique qualities. The lower heating and cooling costs as well as the low-maintenance finish bring ongoing savings.

Australian-Santa Fe elegance in rural Victoria

John Walker and Anne Stelling have owner-built one of Australia's most elegant strawbale homes.

Imagine hiring a Scottish-Estonian architect to design a Santa Fe-style home on 150 acres in rural Australia. That's exactly what Anne Stelling and John Walker did on their land near Beechworth in north-east Victoria, and the result is one of the most stunningly-beautiful homes you could ever hope to see. The three bedroom house is the ultimate in passive solar design, with heaps of north-facing glass doors and windows, pergolas and deciduous grape vines to offer summer shade, a huge, sweeping clerestory window to soak in the winter sun, and the whole lot 'tied together' in a beautiful and functional design.

"When we first went to see our architect, Bernie Jovaras, we showed him our cute little drawing of an octagonal house. He smiled, told us it was very sweet, then put it to one side and came up with this. We love it," says Anne. And with good reason. The house has a peaceful and elegant atmosphere, and the views over the paddocks and wetland are spectacular.

John and Anne started building in November 1997 and with their baby daughter, Maisie, have been in the completed house since January 1999. They are very modest and successful owner builders: their house is completely finished! Often it seems that owner-built homes need lots of small 'finishing-off' jobs, but not so in this house.

The concrete slab floor absorbs the winter sun, and they have hydronic heating via their Australian-made Thermalux wood stove, plus an LP gas stove. The house has solar hot water, an array of 12 X 80-watt solar panels, a composting toilet system, and an energy-efficient, Frostbite 24 volt electric fridge, which uses very little of their precious solar power.

Anne says that their one regret was deciding not to spend the $8,000 it would have cost to have their architect supervise the actual construction phase. Some people think that an architect simply draws up house plans, but hands-on architects who supervise construction, negotiate and supervise contractors and tradespeople, and know how and where to buy unusual materials, can end up saving their clients big money. Anne said that they would have saved the $8,000 through the mistakes that their architect would have avoided.

A clever idea in their concrete slab design was to recess the slab for the living areas to allow for battens and a timber floor which is flush with the height of the slab.

This way, they can still enjoy the warmth and 'give' of a timber floor, without sacrificing the energy efficiency of a concrete slab.

The kitchen is very inviting with recycled redgum benchtops varnished with

John Walker and Anne Stelling's owner-built home is one of Australia's truly outstanding examples of successful strawbale building.

Bio-Products floor varnish, and a huge redgum chopping block the size of a dining room table separating the kitchen from the open plan living and dining area.

The truth window is the most individual and intriguing I've seen. Anne's grandfather manufactured cast iron wood stoves, and for many years she has kept a cast iron firebox door with her family name, 'Stelling' cast in raised letters. The door has found a permanent home, mounted on the wall — opening the door reveals the straw within.

The couple hired local builders to apply three coats of cement render but apart from this cost, they seem to have done nearly all the major building tasks themselves, or with the help of friends and family.

They picked willow branches from the creek about 1.5 metres long and drove them into the bales as pins. They had flat steel plate, the width of their bales and footings, bent at each end and holes drilled so they could dynabolt the plates to the slab and tie back the chicken wire covering their walls to holes in either end of the plates.

The understated elegance of John and Anne's home (on the front cover at bottom right) and their modesty about its charm, left us feeling that this home is one of Australia's truly outstanding examples of successful strawbale building. — *Alan*

Professional owner-building in a cold climate

Andy and Daniela Donaldson and their children are building a 40 square home at Stanley in north-east Victoria.

The Donaldson home is double-storey with a garage 'wing'. Andy is a builder and is constructing everything, including the wooden staircase, the cellar with its timber shelves, and a lovely Tuscan-style wrought iron balcony off the master bedroom.

"We wanted a living house. We can have all the windows closed in the middle of winter, but the house still breathes," says Andy. Stanley has a cold climate so the Donaldsons have created a well-insulated home. They have hydronic floor heating: pipes with hot water circulating through the slab, moved by a 20 watt pump attached to the hot water jacket of their Stanley wood stove. The stove provides all their hot water, central heating and cooking heat, and is really the heart of the house.

The slab is protected from the cold: the sub-floor starts with packing sand, then two layers of black plastic, styrofoam insulation, then light mesh, heating pipes, another sheet of mesh, then three inches (75 mm) of slab. Houses lose about 35 per cent of their heat through the ceiling, and the Donaldsons pumped loose wool insulation 15 cm (6 inches) thick into the ceiling space to get an 'R' value of 4.0.

The windows are all double-glazed. Andy and Daniela have used plantation or recycled timbers, and Stegbar assured Andy that the windows are made from their own American cedar plantations. The downstairs internal walls are pressed earth blocks which absorb the winter sun through the north-facing windows, to release heat energy long after the sun has set. The house has cost around $90,000 and Andy will spend about another $1000 to finish it. Daniela designed the house, Andy drew up the plans.

Structure and render

The house frame is oregon which Andy milled from a plantation established in the 1930s only 5 km away. Andy isn't worried about termites, and has a 100 mm visual barrier on the perimeter of the slab. For the damp-proof course, Andy used three layers of thick bituminous paint straight on the footings, and around the cellar.

"With the first coat, the concrete just sucked it in. The second coat filled in all the holes, and the third coat put a glaze on it!"

The bituminous paint seals any holes around the reo rods protruding from the slab. Andy buys the paint in 60 litre drums for around $20 but it goes hard "so only buy it as you go". The paint can be bought from road patching suppliers. "Look at old techniques — a hundred years ago they always had a tar barrier under their bricks," he says.

Walls and render

The strawbale walls go right to the top of the ridge, rather than having weatherboards infill these high areas. Andy needed odd-shaped bale pieces as he got near the ridge, so he made a beater ("like a big 4 by 2 cricket bat") to beat the bale edge to fit.

Andy didn't tie down the walls — he just left them for six months to settle, and they dropped around 5 cm (2 inches). He also didn't bother covering the walls in

For the damp-proof course, Andy used three layers of thick bituminous paint straight on the footings, and around the cellar.

Andy's external render mix was: 3 parts cement, 12 parts sand, and 1 part hydrated lime.

chicken wire: "I think chicken wire's a waste of time except around doors and windows." The four upstairs bedrooms have stud walls lined in plasterboard, and Andy has infilled the cavities with biscuits of loose straw. The upstairs floor is plantation radiata pine polished with three coats of linseed oil and turpentine. The 6 X 4 metre cellar is made of Besser blocks and is lined with timber racks to store the juicing apples, the bottled preserves, and the home brew that the Donaldsons produce.

Andy seems to have enjoyed rendering the walls. The house has three coats of render, and Andy's external render mix was 3 parts cement, 12 parts sand, and 1 part hydrated lime. He says the secret was to have a batch mixing in the concrete mixer for as long as possible before applying it because this made the mix "fluffy and aerated".

"If it wasn't fluffed up it was really hard to apply. I just bought normal hydrated lime and I had the mixer going all day. While I was applying a batch of render I had the next lot in the mixer. It sticks beautifully and I didn't need chicken wire." Their inside render mix was: 1 part Brightlight cement, 12 parts sand, and 2 parts lime.

The Donaldsons put a little yellow oxide in the final coat so the finished walls have a lovely pale yellow glow that gives the house a happy, snug atmosphere — perfect for their stunning mountain climate.

— *Alan*

The winery's high-pitched building frame is actually made from three steel farm sheds, each 8 X 20 metres.

Grand biodynamic winery uses strawbale to great effect

Julian Castagna and his family are building a huge, 48 square strawbale winery near Beechworth in north-east Victoria. The winery sits on a low hill with commanding views to the north and south of distant ranges of hills. The neat rows of grape vines (successfully made into biodynamic wine under the 'Castagna' label) drape down the hills away from the winery, and the property feels like the grounds of a French chateau.

The high-pitched building frame is actually made from three steel farm sheds, each 8 X 20 metres, and joined in a 'U'-shape to create a massive central courtyard — you can almost here the clip-clop of the horses' hooves as the carriage delivers the master to the chateau!

"I built the place out of *The Straw Bale House* but we were adapting and inventing all the time. For instance we learned the hard way that 6 metre spans are too long — 3 metre spans are the best size. I designed the building first but I learnt that we should consider the materials first before finalising the design. There was no one experienced to help us so we had to invent a lot ourselves as we went along," says Julian.

All the timber is either recycled or plantation pine, and at the moment, one wing of the winery is used as a home and office. Eventually the Castagnas may turn this wing into B&B accommodation. The huge adzed frames of the sliding winery doors are called 'The Sentinels' and they glow with strength, texture and character, offsetting the clay-coloured walls superbly. The walls were all hand-rendered in two coats with a standard cement render applied by a Spanish tradesman.

The floor of the office wing is exquisite terracotta tiles that look like they cost a fortune to import from Italy, except that they are all handmade by the Castagnas from cement, lime, colouring and screened road-making stones. The winery is not finished, but the combination of the Castagnas' artistic flair and great energy are very prominent in the whole project. — *Alan*

TASMANIA

Strawbale house defies Tasmanian winter

Architect Peter Scott designed a house that doesn't need a major heating source.

by Peter Scott
Fern Tree, Tasmania.

WHEN THE TIME came for Keith and Fran Cobbold and their three children to quit renting and build a house of their own they bought about two hectares of land at Howden, south of Hobart and west of the River Derwent, overlooking the North-West Bay and the hills behind Mount Wellington. Their land slopes down towards the south-west, towards the water and away from the sun.

I was approached to investigate ways in which a design could accommodate their requirements for a house which would take advantage of the terrific views across the bay and also provide good passive solar performance. It was not immediately obvious that these two requirements were compatible. However, the discipline of providing a good solution within a short time frame made me focus on the two fundamental issues: how to make the house perform well and how to ensure that it was in harmony with the environment.

Design

Keith and Fran had other elements they wanted incorporated into the plan: they wanted a strawbale house, largely open plan; they wanted an upper-level master bedroom; they wanted the house to be connected to the land visually and physically. My response was to propose a house that was grounded in the site, that hugged the contour of the land and whose single plane of roofing dipped down, seeming almost to touch the ground before rising to accommodate the upper level bedroom.

This design was refined after a series of meetings. I pared back the amount of accommodation to what could be encompassed in the core curved plan which only widened the view from the whole length of the house and accentuated the amount of passive solar gain available in the southern end. It was decided that external decks were unnecessary when the site itself presented one big undulating deck.

Heating

As we scrutinised the way in which the house would work in more detail, it became apparent that the passive solar design and the superb insulation that the strawbale walls offered made the inclusion of a major heat source completely unnecessary — so the cost and space devoted to a wood heater were deleted. Instead, the rising, curving cathedral ceiling inside is utilised to channel warmed air to the apex by convection, from which it is ducted through the outer wall and returned to floor level.

Keith and Fran wanted to include an internal garden and flow forms to invigorate the internal atmosphere. The simple air circulation system complemented the incorporation of this life and movement within the house.

Top: The open-plan interior of the house. The cathedral ceiling channels warmed air to the apex by convection.
Left: Constructing the strawbale walls.
Right: The sculpted walls of the Cobbolds' house: designed to take advantage of magnificent views and provide passive solar performance.

Savings

It is fascinating to see how this house, one of about 20 established, or being established, on this new subdivision and built with care and conviction by the owners, evokes such interest compared to its traditionally constructed neighbours. For a comparable cost to others around them the Cobbolds have built a house which is tailor-made for them, uses recycled and waste materials and which gives them dramatic savings in running costs. Most importantly, it is designed for its site, in the way it performs and how it rests in the landscape. I hope that it proves to many of the people who have followed its construction that strawbale building is a viable and appropriate technology for homes.

With a little help from our (50) friends

If you can buy an ideal block of land half a world away, then building your own house with no experience is an adventure too.

by Nic Goodwolf
Cygnet, Tasmania.

I WAS LIVING in Scotland with my partner, Maddy, when we bought land for sale in Tasmania. Friends were buying the two blocks next to it in a little valley eight kilometres from Cygnet, which is 44 kilometres from Hobart. It sounds crazy to buy land from the other side of the world just by looking at photos and maps and hearing vague descriptions from our friends. I would not recommend it, but in our case it made sense and we have been very lucky.

PLANS

On arriving in Tasmania we rented a house in Cygnet with our new baby boy. We were excited to see our dream of living in our own home soon take shape. We spent a lot of time talking to Bruce, Maddy's Dad who is an architect, and a few weeks later we ended up with a set of drawings we all felt very happy about. Bruce measured the slight slope of the land, and he was very conscious that we wanted our house to have only a minimal impact on the land. So our house has five different levels, often there is only one step difference between one level and the next. This careful planning has avoided the scars of embankments that are often seen around houses built on a slope.

There was a lot of talk about strawbale building in Scotland so we thought that this was our chance to try it out. As our house was the first strawbale building that the Huon Council had dealt with, we decided not to push them too far and we limited our strawbaling to three walls which were all infill and had no windows and doors in them.

After many letters and the realisation that there was very little real information about strawbale being used in Tasmania, the building inspector agreed to use our dwelling as a kind of 'see as we go' test case for the council.

GETTING STARTED

At this time we were juggling my final year at art school, our son, the hoarding of salvaged materials and weekly trips to auctions and the tip. We were also starting the first stages of building. I had not properly built anything before which didn't make it easier, but at least I didn't know what we were in for.

The soil tester came and inspected the ground. It became clear, very quickly, that we require very deep foundations due to our very inconsistent soil structure. It did not take a genius to realise how many more thousands of dollars it would have cost us if we proceeded with the plan of building all walls except the north wall out of strawbale.

In retrospect I am very glad about

The strawbale walls that Nic built, he found rendering the hardest part.

building a strawbale—stud frame hybrid. We still ended up with the lovely feeling of the bale walls in the livingroom, and we gained a lot of space which would have been taken up by the bale walls in the rest of the building.

We poured the slab in which I inserted half-inch poly pipe. Perhaps one day I will hook it up to the hot water and it will be just as nice as the floor heating I remembered from my childhood in Germany. Slab and ring footings were done with a lot of help from friends. All the formwork reinforcements, pouring and the red staining of the slab was very labour intensive and we all needed a good break after our first big step.

Building small

Wil, Maddy's brother who is a lovely carpenter, came to help me for two weeks to erect the frame. It works so much faster when you know what you are doing. I didn't know what I was doing but Wil did. If you decide to pay anyone during the building stage, I would advise that you choose these two weeks.

Our house was very small back then. We only built the kitchen, livingroom and bathroom. We knew if we started small that we could manage the building better, we could afford to build it and the extensions that you build later are easy to do if you are already living comfortably on your land.

The strawbale walls followed pretty quickly. The books that I had read on strawbale building did not prepare me for the construction. Perhaps I am the learning-by-doing type. We were daunted by the sheer size of the walls when it came to rendering. Stacking bales four metres high was still alright, but trying to render that high, on wobbly scaffolding, was another thing.

We started using long 'chemical gloves'

Strawbale walls have a sculpted feel in the interior of Nic and Maddy's home.

from the hardware shop for the first coat. Everything else was too frustrating. But in the end we had developed our own technique and we felt we had it in control. The brown coat went up and we were happy. We knew that at least now the walls were more or less water-proof and we started doing more work inside the house.

FINISHING OFF

All our windows, with a few exceptions, were built by us, which took quite some time but it was worthwhile doing. I searched all over town for glass that was a decent size. I bought bullet-proof glass from banks that were closing down; glass shelves from old shop fittings, and recycled glass. This enabled us to have floor to ceiling glass on all northern walls in the building, and as the glass is up to ten millimetres thick we had no problem with the building regulations.

Time was running out as we had set Christmas as our deadline. We quickly finished the inside bale walls and painted them, did the plastering on the stud walls, had the electrician come and to everyone's surprise we moved in three days before Christmas Day.

Our house took us 12 months to build. It was built with the help of 50 people all of whom gave their help free, except the plumber, electrician and digger and others who delivered materials. Thank you so much! We could not have done it without you.

EXTENSIONS

All this happened just over three years ago. We still love our house. Gus, our son, is happy to have his own room so he does not have to sleep on the bathroom floor anymore. We are happy to have our own room too, and I am wondering what we will do when our second child arrives.

All bale walls have been finished in a dark red cement render and the timber cladding is stained black. I started giving adult education courses recently on strawbale building.

It seems I can't get away from strawbale building. The courses have been a great success. The students love to experience building firsthand — it is good when you realise that you aren't the only one who is learning by doing.

And on the seventh year they rested

The first council-approved strawbale house in the state took six years all up and is now a comfortable, four-bedroom home.

by Danny Vonk
Molesworth, Tasmania.

AFTER A 12 month tour around Australia in the great Outback we felt cramped when we came home to suburbia. So my wife, Jo-ann, and I bought an owner-built home on three hectares of bush, just 30 minutes from Hobart. Molesworth is in the Derwent Valley with a great community and plenty of owner-builders and alternative style people. Life was good until our new home burnt down, which was caused by a faulty flue on the wood heater.

We had to rent again. We decided to build another house and because I'm a builder and plumber we thought that we'd build big because we're saving on labour costs. We designed a home using the sandstone which is found on our property. After we gained the necessary approval we started building stone walls. Three years and a lot of sweat and blistered hands later, we got to the floor level. At this rate we would finish when I was 80, so we went back to the drawing board. A friend

Danny selects the driest bales for placing into position in the wall.

handed me a book on strawbale. And I thought, hey, this isn't so bad.

So off we went to the council engineer. He'd seen this type of building in the USA and he sent a letter requesting strawbale building guidelines from the State of California. We then redesigned the same plan, changing from sandstone to post-and-beam with strawbale infill rendered over with a final coat that has a sandstone oxide and textured finish.

Off we went to council with our revised plans, but this time they wouldn't approve them on the grounds that they had insufficient information. After much deliberation the council, who were very helpful, drew up their own strawbale construction guidelines for infill walls (non-loadbearing) and we gained approval.

We put a frame up on the slab. All internal walls were standard stud and plaster. The door frame was traditionally pitched. The open-plan loungeroom, diningroom and kitchen have a cathedral ceiling. With the frame complete we were ready to infill with strawbales.

BALES

Straw is a general term given to the dry stems of cereal grains left after the seed heads have been removed and processed. Cereal crops in Australia are wheat, rye, barley, oats and rice. The most popular of these is wheat, which is grown Australia-wide and the straw from wheat is readily available. Once the heads have been removed and threshed the remainder of the plant is left on the ground. The straw is then collected by the bailer, compressed into rectangular bales and bound by twine or bailing wire.

The best and cheapest time to buy bales is at harvest time. When you have decided on what type of straw to buy then ring around to find who is growing that crop and when they are going to harvest.

When inquiring, explain what you are using the bales for and that you require the bales to be specially made for you. Bales must be tight, as tight as the machine can go without breaking the twine or wire. Bales must be as dry as possible. Twenty per cent or less moisture content is ideal. Bales left on the ground exposed to the elements are not good enough. Try to arrange pick up and storage on the day of harvesting.

If picking up bales yourself, then wear gloves, a long-sleeved shirt and long trousers and boots because strawbales will scratch and cut your skin.

To my knowledge there are different types of bailing machines and different sized bales. The bales we used are 1000 millimetres long by 500 millimetres wide by 400 millimetres high. The bales are used by laying them flat so that the largest cross-section is horizontal. When calculating the amount of bales for your particular job the calculation is 2.5 bales per square metre of wall. For example: 12 by 2.4 = 28.8 square metre wall, 28.8 by 2.5 = 72 bales. But always allow for more bales because of damage done due to transport, storage and handling.

Moisture is the largest problem with strawbales. Avoid moisture at all costs because it could mean having to start the wall again. I'm a plumber by trade so I devised a way to damp-course proof and prevent moisture ingress through capillary attraction. This is what I did.

WALLS

We began the wall with a metal flashing on a step slab around all walls to be strawbaled. On the flashing we fastened t-plates to support poles. Where there are doorways we turned up the flashing to create a box gutter where the bales will go. Nylon tappins were used to fasten these to the concrete and all joints, and we sealed all joints and tappins with silicone sealer.

We set out the position of bales and marked for starter pins two per bale. We drilled through flashing and into concrete as far as the drill will go using a 12 millimetre Y bar — about 200 millimetres. We cut reobars 800 millimetres long and bashed them in. The rod is be long enough to reach into the second bale.

We rolled out 100 millimetres of black Agroflex pipe and placed it on the lower section of flashing and we kept it in posi-

The final coat of render was done by a professional plasterer to get the required finish.

Right: A bale goes in next to a window frame. Danny made all the window and door frames.

tion with saddles. This keeps the first bale up off the lower slab and prevents water from penetrating. We started laying by impaling bales onto bars and bashing, kicking, whatever, to keep them roughly in line. The second row will cover the starter bars. Lay bales like bricks: for this you need to make halfbales or bales to a particular shape.

We made a large needle that was 700 millimetres long. We flattened one end on an anvil and drilled four to five holes very close together then we redrilled holes at 45 degrees making a slot. We cleaned it up with a chainsaw file.

We used baling twine (buy a three kilometre roll, you'll use it) and stitched the bales as follows. Feed the other end through a loop and pull it tight like a truckie's knot and tie off, then cut excess away. Tighten to a point when the original string becomes slightly slack then tie off. You can do it to make two small bales out of one or a stepped bale to go around a pole.

A bit of experimentation is required here to get the hang of it. Make sure to keep the bits that are left over dry because you'll need them to pack out and fill holes with later. And it makes an excellent mulch if you don't use it.

We laid a section of wall to the top rail then we made spikes to pin all the bales right through and we attached these to the rail with nails. We then put on the roof to protect it. If the roof is already on you'll have to spike the rows as you go. How do you fix the spiked top bales to the rail at the top? Some friends of ours found it extremely difficult as they had no room but needed space to keep bales dry. We were fortunate as a neighbour offered a storage shed to keep the bales and we brought over what we needed for each section.

We started early with as many hands as possible. Each person was given a particular task, tidying, laying, spikes, getting bales, for example, and by the afternoon we were onto the roof, after spikes were put through and fixed.

Doors and windows

We made our own door and window frames. The inside frame is traditional and has a sloped rebate at the bottom and rebate all around for openers and glass. The outside has a 50 millimetre by 50 millimetre section and we attached the centre to that. The vertical sides were like studs to locate the frame, for support, and to generally make it easier and draught-proof. We fastened the side stud to the top timber rail with screws.

We then laid bales up to the studs, leaving a 50 millimetre gap between the bale and frame to be filled with render later. When we got to the last row of bales we had to cut a slot along an entire length to get around the top rail. To achieve this we used a chainsaw like a router, we also used the chainsaw for numerous other odd shaped bales and for trimming the bales. But if you are using a chainsaw then avoid the strings or the bale will spring apart.

Wire

With all bales in and the roof complete we were ready to wire. We covered all walls inside and out with chicken-wire. It comes in 100 metre rolls and we used six rolls. We started from the bottom roll out along the wall. We put in temporary pins using made-up staples out of ten-gauge wire to keep it in place.

Then we cut and bent the wire to the windows and attached it to the timber surround around the window frames. We attached the wire with builders' strap and screws. At the window sill we placed a sheet of plastic 200 millimetres under the wire and sloped slightly to prevent rainwater soaking through the sill and into the bales. At the window head we doubled up the wire using expanded metal lath. We also used this on external corners and doorways for increased strength.

We over-lapped each row of wire about 100 millimetres. We also packed out and trimmed up each section of wire with loose straw to obtain a fairly even line. The wire will give you roughly the final shape so it's important to obtain your desired shape at this time. Then we made special staples out of ten-gauge wire to tie the inside wire to the outside wire. These should be pulled tight then bent up and bent back into the bale. At the top-rail bend the wire up then staple, screw, and nail it.

Rendering

The wire is the reinforcing for the render: when rendered the wall is like a vertical slab. The strawbales have no structural strength, they are just infill with excellent thermal properties — all the strength is in the render and wire.

We rendered in three coats: the first two were scratch coats with the final coat being coloured and wiped with a foam pad to create a soft look. All corners were rounded. The inside was sealed and painted using acrylic paints. The render must not be water-proofed, it must breathe to allow any moisture to dissipate. We cheated with rendering and paid a professional solid plasterer to complete the render to our specification.

Final points

Mice were a problem during construction. When the building was completed we finally used poison to get rid of them. Now, we're in the house it's beautiful. It's beautiful and cool in summer and in winter it's warm.

Constructive building

Building an office and workshop was the first step to a growing commitment to strawbale building.

by Nik Vollmer
Kingston, Tasmania.

I FIRST SAW a strawbale construction at the Sydney Home Show, held at Darling Harbour, and it immediately took my interest. I am a carpenter and did not want to only build conventional brick and tile homes. I found the idea of building with straw exciting as it fits in with Tasmania's natural environment and it is environmentally friendly.

The first step after seeing the display in Sydney was to contact our passive solar designer, Nigel Jones, who suggested that we build an office and workshop. We now use it as a display for interested people who are then able to experience the sense of calm and peace that strawbale construction gives.

I worked closely with Nigel to develop specifications suitable to local councils and Tasmania's weather conditions. We decided on a post and beam construction which enabled us to put the roof on first to protect the bales from the rain.

Building

We excavated a concrete slab with pier holes built into the footings to support the steel columns. On top of each steel column perimeter beams were bolted on and the roofing timbers were pitched onto the beams. Chicken-wire was laid onto the foundations and concrete blocks were laid flat to elevate the bales 100 millimetres off the slab to help protect the bales from moisture.

Rebars were drilled between every second concrete block to pierce the first row of bales to initiate a straight line. A plastic moisture barrier is pierced onto the rebars followed by the first course of strawbales.

Barbed wire is tensioned between each course of bales around the steel columns, to keep them in a straight line. We put two wire ties per bale across the full width of each bale which ties the chicken-wire hard against the bales. Next we knocked in two pegs per bale to hold the barbed wire down hard onto the bales to keep them in a straight line. The bales are laid in the same uniform fashion as bricks.

For the larger windows and doorways leading through the strawbale home a frame is attached from the perimeter beam to the slab and for smaller windows a frame is built oversize of the window to form an opening which 'floats' within the bales.

Chicken-wire was attached to the bales stretching from the slab to the perimeter beam and wire ties were used to fix the chicken-wire tightly to the bales. This strengthens and stabilises the walls. A large block of wood was used to flatten out any uneven bulges in the walls.

All windows and doors were then fixed into position which left only the rendering to be completed. It was necessary throughout the construction period to keep the bales dry at all times and this was

This office and workshop built of straw has persuaded others to build with bales.

achieved by fixing tarps to the outer side of the exterior walls.

RENDERING

The first coat of render used eight parts sand to one part cement to two parts lime with loose straw mixed in to reinforce the mix. The render was applied using a gloved hand and the render was forced into the bale, filling all cavities and gaps.

Once the render dried the second coat was applied using a similar mix, excluding the loose straw, using a trowel and a wet sponge to smooth off the surface. The end result proved to be a great success as the workshop created a warm, inviting and tranquil display of strawbale construction that has inspired others to try strawbale.

There is so much interest in strawbale building that it will eventually become as popular and accepted as a brick or timber home — with the added benefit of being good for the environment. People are able to become involved in the construction of their own home and that can give a huge amount of satisfaction.

BUT HOW DO I . . . ?

Getting house plans approved

As one who pioneered strawbale house plans through one council, architect Peter Scott can suggest ways to get official approval.

by Peter Scott
Fern Tree, Tasmania.

THE HOUSE I designed for the Cobbold family at Howden was the first strawbale house to go through the full development, building and plumbing application process under Kingborough Council in southern Tasmania. It acted as a test case and has helped to establish the guidelines that the Kingborough Council use to assess other strawbale constructions in the municipality.

This fully approved project also demonstrated that strawbale building, although in this case an 'alternative solution' in some respects under the Building Code of Australia (BCA) does not have to be, and should not have to be, undertaken out of the eye of the local statutory authority.

APPROACHING COUNCIL

For this project it was an advantage to have a good working relationship with the officers of Kingborough Council which I have developed in my professional capacity as an architect. It would be equally possible and valuable for owner-builders to foster good relations with their local authority.

A good way of developing good relations with council officials, especially for innovative projects, is to keep the council officers informed of your intentions before you make a formal application. Then they have the opportunity to offer advice and have a sense of involvement in the project which can help to smooth the approval process.

And, where councils have considered previous strawbale building applications, their advice about requirements and presentation of information may be helpful. They may also be able to locate other helpful sources of information. Kingborough Council have started to build up a reference library of relevant technical material to help them assess similar applications: this information could also be available to potential applicants.

ENGINEERS

A considerable part of council approval regards building structure, and engineering certification is the key to obtaining this approval. Engineering certification is usually accepted, and should be accepted, as sufficient evidence that the design of the building structure is sound.

On this project I prepared all the working and engineering drawings in tandem with an engineer with whom I have worked regularly, and we discussed construction details before they were finalised in a form that he was happy to certify.

An aid to us in this regard was the decision, made for aesthetic reasons, but of great benefit in achieving a straightforward approval process, that this house be built with a loadbearing timber pole frame and not with loadbearing strawbale walls.

While several aspects of the timber frame fall outside the 'deemed to satisfy' requirements of the BCA, for which specific engineering calculations were undertaken, most aspects fall within areas covered by Australian Standard 1684 (National Timber Framing Code) and the BCA. This makes certification by an engineer and council assessment much simpler and therefore more certain, quicker and cheaper.

Where more detailed engineering input was needed, for example, where the paucity of solid structure took wind bracing beyond the realm of AS 1684 and into the realm of an 'alternative solution' under the BCA, the engineer calculated specific wind loads and we were then able to discuss how to incorporate sufficient bracing.

In this way a fully certified set of documents were provided to the council and a major part of the information requiring approval was taken care of.

It is worth noting that some engineers are willing to undertake calculations for, and certify, loadbearing strawbale structures, and some councils may be willing to accept precedents as adequate demonstration of an 'alternative solution' under the BCA.

In both cases I would anticipate a longer and probably more expensive certification and approval process, which should be borne in mind when considering such an approach.

Drawings

I prepared all the documents for this project and always take care to provide a clear set of drawings — including a high level of detailing. This provides builders with clear directions but also has the following advantages:

- By thinking through the details of construction in some thoroughness at the design stage, it is possible to address issues which council will be considering before granting an approval.

For example, finishes to the floor junction and face of the strawbale wall in the bathroom of this house were specified to meet the water protection requirements of the BCA. By covering bases in this way approval was achieved on this project without need for supplementary information of any kind.

- A good set of drawings gives the council confidence that, even where there is an uncertainty about a particular aspect of the construction, it will be addressed and dealt with in a similarly professional manner.

Finding precedents

Because I discussed council's likely requirements before the building application and I had identified areas where they could not rely on the BCA as the basis for approval, I was then able to provide information on similar construction techniques to those proposed in this house as precedents.

Books on strawbale building, especially *The Straw Bale House*, the Internet and other previously approved projects provided a database from which we were able to identify material which was acceptable to the council as a basis for approval.

I have found councils generally very receptive to uncommon building approaches and open to 'alternative solutions' to the requirements of the BCA. Where full information is provided, and a good working relationship established with the relevant council there should be no reason why applications for strawbale buildings should not be assessed as readily as those for more traditional constructions.

In the foreseeable future there will be Australian standards and a part of the BCA devoted to strawbale construction so that it will be considered under 'deemed to satisfy' criteria. This will truly represent its arrival in the building mainstream.

A council's perspective on strawbale houses

by Nillumbik Environmental Building Surveyors

Nillumbik Shire Council, Victoria.

NILLUMBIK Shire is located just 25 kilometres from the centre of Melbourne. It has the Yarra River as its southern boundary and extends 29 kilometres to Kinglake National Park in the north. The shire covers a total area of 430 square kilometres and has a population of 57,000.

Its people live in close-knit communities ranging from typical urban settings to remote and tranquil bush properties. The shire is a collection of townships each with their own identity and heritage, set amongst bushland and rolling hills. The mudbrick buildings, pioneer cottages and country-style pubs complement the area's natural landscape, while artists' colonies such as Montsalvat and Dunmoochin, and the diverse range of festivals, craft markets and exhibitions all add to the shire's reputation as a centre for creative pursuits.

The shire prides itself as a 'Green wedge' due to the retention of its rural character and preservation of the natural environment.

Residents have the opportunity to live in a relaxed and natural setting in close proximity to a major city. Council's Planning Scheme seeks to enshrine an innovative, sensitive and flexible approach to planning.

The shire seeks to ensure that new development, whether private or public, enhances neighbourhood character and preserves the area's environmental values.

BUILDING TRADITIONS

For more than half a century, mudbrick houses in Nillumbik Shire have enhanced the local environment with their low visual impact and use of natural materials. The artistic community that settled in Eltham in the 1930s brought with them a creative and unconventional spirit and innovative ideas about house design.

Daring to be different, the former Eltham Shire (most of which is included in modern-day Nillumbik) encouraged a range of building methods. Many Eltham builders adopted the adobe construction method during a conservative era more noted for a preponderance of brick veneer and timber buildings.

Today, these adobe homes are an enduring symbol of the area's creative past. There are now an estimated 2000 mudbrick buildings in the shire, along with an emerging trend towards other alternative building materials such as strawbale, stone, steel and pole frame construction.

Building on this tradition, Nillumbik Shire Council has implemented urban design policies that seek to minimise environmental impact through the use of alternative building materials. To facilitate this goal, the shire runs educational and promotional programs to market the latest developments in alternative housing design and construction. Indeed, the shire hosts these programs at venues such as the Eltham Library and the Eltham Community Centre, both of which are monuments to the alternative building philosophy.

LEADING PRACTITIONERS

In support of Council's planning approach,

its building business unit — the Nillumbik Environmental Building Surveyors (NEBS) — have established themselves as leading practitioners in the use of alternative building materials. NEBS is responsible for providing the council's building permits and an inspection service.

It started investigating strawbale housing after receiving inquiries about their use as an alternative building material. Using the Internet, the unit learned that strawbales were first introduced in Europe over 200 years ago. They were then used in Nebraska in the USA during the 1890s in response to a shortage of trees for timber. Many buildings, including houses, farms, churches and schools, applied this material. But it wasn't until the 1970s that there was a rapid expansion of the material throughout America and Canada.

Usage of strawbales as a building material has generally been met with scepticism by building surveyors, whether employed by local government or in the private sector. Many building surveyors are reluctant to approve permits for this type of construction because they are either inexperienced in its use or resistant to new building techniques.

New agenda

Nillumbik, with its experience in adobe construction, has a very different attitude. Council and NEBS are setting an exciting new agenda for housing construction using strawbale and, in doing so, breaking new ground in local government building surveying.

NEBS building inspector, Chas McKinna, has studied strawbale construction at Holmesglen TAFE. He says anyone contemplating building in strawbales should enrol in a similar course, and especially one that includes a hands-on component.

Council's approach to strawbale housing is akin to its approach to mudbrick housing. Drawing on their decades of experience with adobe construction, NEBS are applying the same design and construction principles. Strawbales are regarded basically as oversized mudbricks, and the same building principles are applied to ensure loads are supported by the structural frame and joints are weatherproofed.

Non-loadbearing strawbale construction is treated much the same as non-loadbearing mudbrick building, where the major roof loads are supported by columns and beams. Loadbearing construction, where the major roof loads are supported directly onto strawbale walls, is still being researched by NEBS.

Significant issues

The significant issues pertaining to strawbale construction that all builders should consider are:
• settlement of the strawbales;
• compression of the strawbales;
• bale density and the consistency of the density;
• moisture content of the bales;
• the loadbearing capability of strawbales as opposed to non-loadbearing strawbales;
• termite barriers;
• construction in bushfire areas;
• water-proofing of the strawbales — external and internal renders;
• fixing of the strawbales;
• placement and water-proofing of plumbing fixtures adjacent to strawbale walls.

Advantages of strawbale

The advantages of strawbale construction include:
• excellent insulation properties resulting in considerable energy reduction and lower heating and cooling costs;
• ease of construction — you don't have to be a technical genius to lay strawbales;
• the cost of strawbales is low compared to other building materials.

Disadvantages of strawbale

On the other hand, a strawbale builder can be disadvantaged by having to pay increased construction costs. For example, you may need to use additional concrete (both for strip footings or a concrete slab) to accommodate additional wall thickness.

Strawbale construction is very labour intensive as you will need to cut, lift and render the bales.

This means that while strawbales are very cheap, costs will mount if you have to employ someone to build your home.

STRAWBALE AND THE BCA

The Building Code of Australia (BCA) is generally divided into two sections: Performance Provisions and Acceptable Construction.

Conventional forms of construction, such as brick or weatherboard are normally supported by Australian Standards and come under the Acceptable Construction category. (It is worth noting that adobe construction generally falls into this same category in accordance with a CSIRO document.)

But strawbales are not a 'deemed to comply' material like brick or weatherboard and are therefore considered under the Performance Provisions.

The Performance Provisions category contains objectives, functional statements and performance requirements that need to be considered in an application assessment.

For example, the provision that relates to dampness and weather-proofing provides for the following and is applicable to strawbale construction:

The objective is to:

1. safeguard occupants from illness or injury and protect the building from damage caused by:
• surface water;
• external moisture entering a building;
• the accumulation of internal moisture in a building;
• discharge of swimming pool waste water.

2. protect other property from damage caused by:
• redirected surface water;
• the discharge of swimming pool waste water.

The functional statement is: "A building is to be constructed to provide resistance to moisture from the outside and moisture rising from the ground."

The performance requirement is:
Weather-proofing:
A roof and external wall (including openings around windows and doors) must prevent the penetration of water that could cause:
• unhealthy or dangerous conditions, or loss of amenity for occupants;
• undue dampness or deterioration of building elements.

Dampness:
Moisture from the ground must be prevented from causing:
• unhealthy or dangerous conditions, or loss of amenity for occupants;
• undue dampness or deterioration of building elements.

LOOKING FORWARD

In building permits issued to date, NEBS has had to address these and other issues. Because the main structure in most strawbale homes is supported by posts and beams with the strawbale walls being infill panels between the posts, approval is relatively straightforward.

The main structure can be designed utilising standard engineering principles or timber framing manuals depending on the spans and spacing of the framing members. NEBS is also considering other uses and designs for strawbale construction and how best to assist in the approval of these.

Strawbale homes represent an exciting and challenging form of construction not just in the Nillumbik Shire but across Australia. We are proud to be associated with this innovative and exciting new construction type.

Life beyond brick veneer

How an engineer got hooked on designing and engineering strawbale houses — 20 so far!

by Michael Kingsbury
Conondale, Queensland.

IT ALL BEGAN when I read about strawbales. This made an immediate impact on a very disgruntled civil engineer who had recently turned his back on the whole construction scene. All of a sudden there was life beyond the brick veneer boundary that didn't involve digging holes and moulding tonnes of earth to build a wall; I was truly impressed, at last.

So I went to Tyalgum, the permaculture capital, which was like going home because I had lived at Tyalgum when I was an engineer with the Tweed Shire Council.

For four days, I was into straw and mud (it rained a lot), and loving it! With a bunch of balers we stacked strawbales, spliced them and then rendered three walls of a rabbit shed. By the end of the first week, I'd scored a design job using strawbales, and then got it approved by Lismore City Council. I'd arrived!

Since that time I have designed numerous council approved buildings. I have also conducted workshops and given talks to potential strawbale house owner-builders, and these talks got a great response. There is a great interest in this type of construction, and with good reason.

Approval

My greatest moment was receiving approval for the first strawbale house design that I lodged in Queensland (with Caloundra City Council). It took three months to satisfy the council until, to their credit, the building department, then the council finally approved the application.

The house was a series of three interconnected octagonal buildings, incorporating loadbearing, non-loadbearing and hybrid components, pole-framed construction, with loft and belfry, part concrete floor and part timber floor. Ambitious? Definitely! Impossible? Never! The shame of it all is that the house was never built: I sold the land and moved into an old Queenslander which I have been renovating ever since.

When the first of the strawbale houses I had designed was under construction, the entire building section came out for a council inspection. I have continued with my engineering practice, providing designs, advice or building assistance for those choosing strawbales. There are so many good points to building with strawbales, and construction can be an adventure.

Versatile material

Which other material can you kick, punch, shape with a whipper snipper, tie up, and have a good time with? With a well-placed nudge you can shape each bale into a curve for free forming walls or fences, mould fireplaces, and build 'built in' chairs and tables. The bales make excellent scaffolding during the building process and the scraps don't get thrown into a skip when you've finished building, you just mulch a garden bed with them and the soil benefits.

What more could one ask for? Strawbale buildings have character, strength, solidity, longevity, and old world style, along with an insulation value that is second to none (in its simplicity and environmental aspects). Strawbale buildings look and feel great! And let's face it, you don't have to be superman or superwoman to stack strawbales. All you have to be is an individual — just like the building you create. So let's bale it, stack it, render it, roof it, and move in. And above all, let's have fun! Happy baling!

Green house built of straw

Architect Marcus Ward has designed a strawbale house that meets high environmental standards.

by Marcus Ward
Kyneton, Victoria.

THE Verity and McDonald house proposed at Fryers Forest, near Castlemaine in central Victoria, is an interesting response to a very interesting land development. The subdivision development, designed by Holmgren Design Services, is described as a 'forest village' where the principles of permaculture are central to the planning and ongoing use of this 210 hectare project. Eleven shareholders own an allotment of around 0.4 hectares (one acre) each. The balance of the property is jointly owned and managed as an experimental agroforestry project.

Comprehensive prescriptions have been made on the design of the houses and their siting in this forest village. One prescription is that houses must generally comply with the Australian Conservation Foundation's Green home guidelines. Unlike typical standards that consider primarily expected energy use alone, the ACF guidelines take a wider view including low-impact material use, impact on infrastructure, and other offsite implications.

A passive solar strawbale house fits the stringent demands of the situation ideally. It is expected to perform brilliantly in terms of energy use with the insulating properties of the external walls combined with the extensive thermal mass of the internal mudbrick walls and concrete slab. It also fits the other Green home criteria by being an extremely low-impact building.

Due primarily to some creative foresight from the David Holmgren team, and people like my clients, the project is a terrific example of what is possible in rural Australia for no more money than a standard brick veneer on a quarter of an acre.

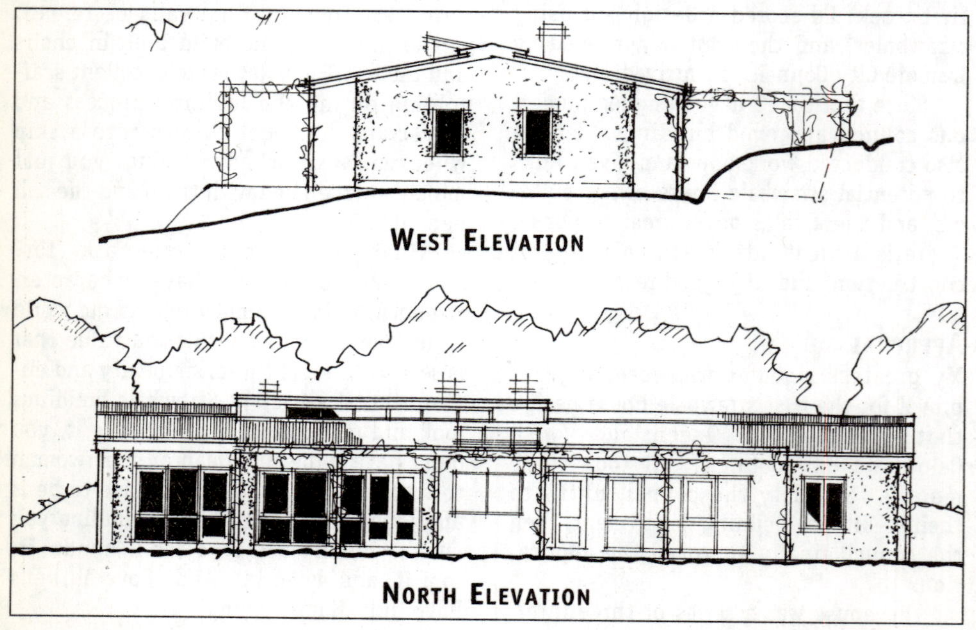

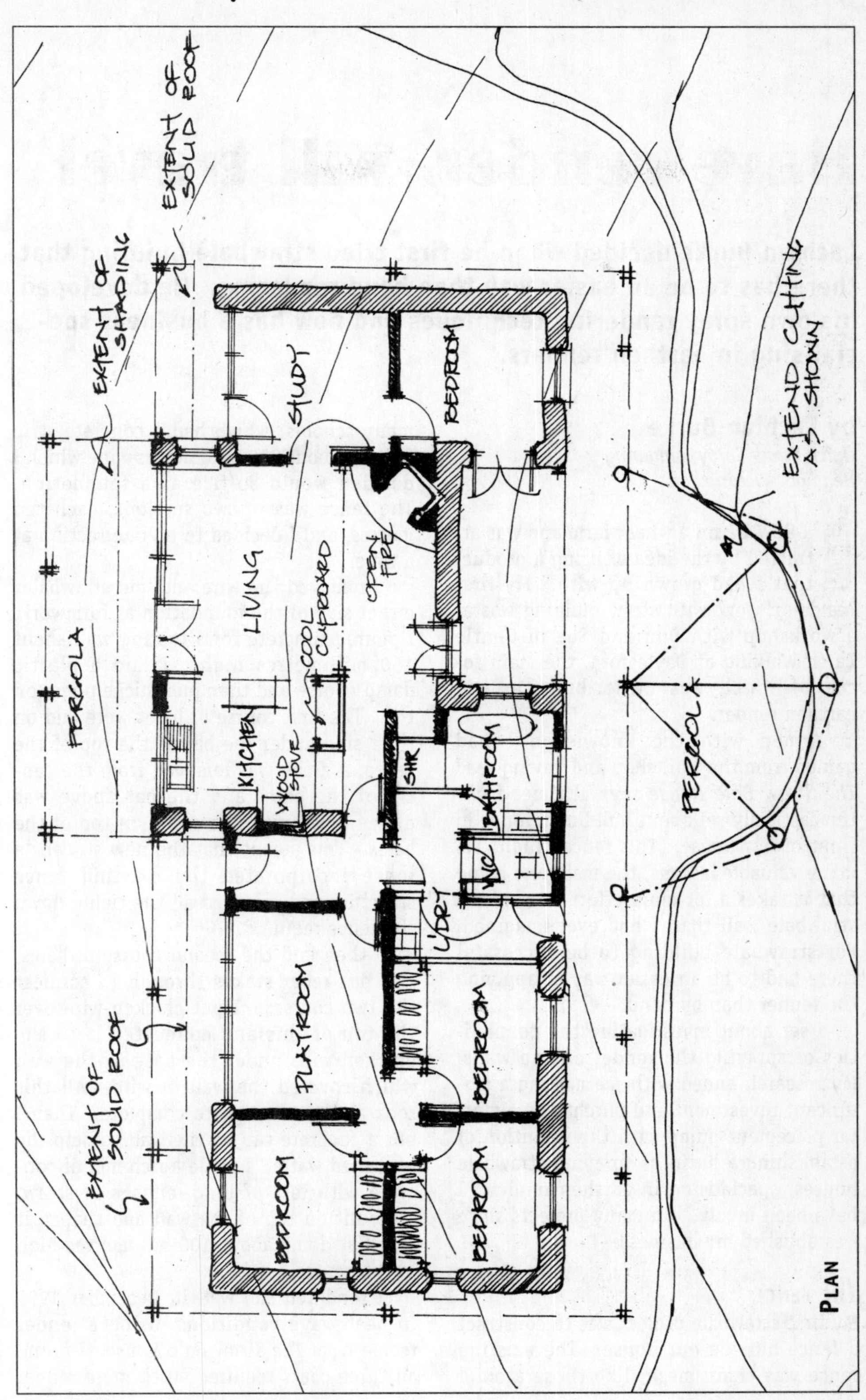
EXTENT OF SOLID ROOF
EXTENT OF SHADING
PERGOLA
STUDY
BEDROOM
LIVING
COOL CUPBOARD
OPEN FIRE
KITCHEN
WOOD STOVE
SHR
BATHROOM
WC
L'DRY
PLAYROOM
BEDROOM
BEDROOM
BEDROOM
EXTENT OF SOLID ROOF
PERGOLA
PLAN

Have render, will travel

Lachlan Burke decided when he first tried strawbale building that there has to be an easier way than hand rendering. He developed his own spray rendering techniques and now has a business specialising in earthen renders.

by Lachlan Burke
Quick-Straw Spray Rendering
Bendigo, Victoria.

I GREW UP on a wheat farm and was attracted by the idea of using a product that I had grown up with. My first hands-on work with straw building was at a workshop with Don and Sue of Gentle Earth Walking at Daylesford, the main focus of which was owner-building and earthen render.

Armed with the knowledge I had gained from the workshop and having read *The Straw Bale House* several times I enthusiastically set about building a fence in front of our house. This fence taught me many valuable lessons, the main one being that it takes a lot more effort to render a strawbale wall than I had ever imagined. For strawbale building to be successful there had to be an easier way of applying the render than by hand.

I set about investigating the possibilities of spraying the render on the walls. My research ended with me making a significant investment and purchasing a mortar placement pump with the intention of establishing a business spraying strawbale houses, specialising in earthen render. I have been involved in many projects since I established my business.

THE FENCE

My first strawbale project was to construct a fence outside our house. The existing fence was chain mesh, like those around many schools, which had a concrete strip footing about 300 millimetres wide which I decided would suffice as a foundation. The fence was in two sections, each ten metres, and I decided to do one section at a time.

I removed the wire and laid strawbales either side of the foundation as formwork. I poured concrete for a footing wall about 150 millimetres high. I laid a plastic damp-course and then put chicken-wire on top. The first course of bales were laid on their side under the bar at the top of the fence, a niche was removed from the centre of the bales and the bar above was driven down into this niche on top of the bales. This meant that the new strawbale fence incorporated the existing fence structure and provided all the tieing down the fence required.

I then laid the second course of bales, and put rebar stakes through to connect the two courses. I put chicken-wire over the top of this and connected it to the chicken-wire under the base of the wall which covered the wall in wire. All this took about two days to complete. Then I put a concrete cap on the wall to help the wall shed water. I made a rich mix of concrete with lots of long lengths of straw, placed it on top of the wall and shaped it to a flat dome about 100 millimetres high in the centre.

I rendered the wall in December 1997 in heat wave conditions, using a render recipe from *The Straw Bale House*. I found all three coats required much more render

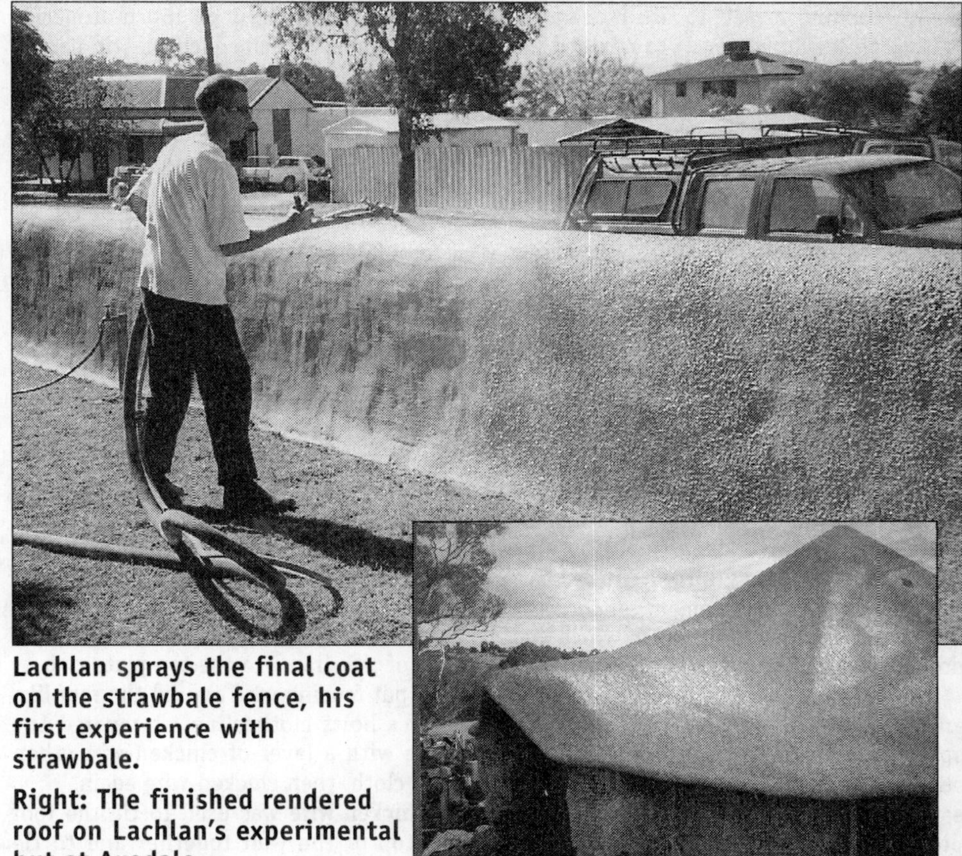

Lachlan sprays the final coat on the strawbale fence, his first experience with strawbale.

Right: The finished rendered roof on Lachlan's experimental hut at Axedale.

than I had envisaged to get a satisfactory coverage and it was more difficult to apply than I thought it would be. The scratch coat was very difficult to push through the wire into the straw and just when I thought I was going well my trowel would catch on the wire and pull large areas of applied render off the wall.

The second and third coats of render were easier to apply than the first coat but it continued to be very hard work and very time consuming. I found that many fine cracks formed on the final coat which meant that I had to rework the surface many times with a trowel and a broom dipped in a render slurry that had the consistency of thick paint.

I think these cracks were caused because the weather was too hot to be rendering in, especially with no shade cover. I also used brickies' sand which was far too fine and contained a clay content that was too high for a cement render. The ideal sand for this job would have been a coarse plasterers' sand which has a lower clay content (about five per cent) and aggregate, starting at about three to four millimetres down to fine.

Spray render machine

The main thing I learnt from this job was that strawbale work was not just a matter of stacking up a few bales and throwing render at them. I was not completely daunted by my first strawbale experience, and I set about investigating a better way of applying render to a bale structure. The obvious solution was to have a machine that could mix and spray the render onto the wall. I found a machine, that was manufactured by a Sydney engineering firm, that I felt could do the job. I set about teaching myself to use it, concentrating on teaching myself to mix and apply earthen renders as this is my preferred method for strawbale houses.

I set about doing some market research on the viability of a business in the strawbale industry aimed at assisting owner-builders. I got a stall at the Seymour Alternative Farm Expo to assess the amount of interest in strawbale construction. It was here that I constructed my second wall and spent three days talking to people about strawbale construction. I came away from Seymour very inspired and felt that it was worth a punt.

FIRST JOB

My first job was to construct two outside seats for the arts department at the Footscray Secondary College. These seats had a straw reinforced cement render. The chairs then had a mosaic finish which was done by the arts students.

I then returned to Don and Sue's place and experimented with earthen render by spraying a small loadbearing shed. We successfully sprayed three coats of an earthen render in three days and had few problems.

The main thing I learnt from this job was that it is still necessary to trowel, or broom, off between coats, even with a machine to blow the render on. This helps fill out the walls and soften any peaks or troughs in the straw construction. A great advantage of spray rendering is that it gives an excellent penetration of render into the bale ends.

EXPERIMENTAL HUT

The next project I completed was an experimental strawbale hut on my property at Axedale, in central Victoria. These walls are loadbearing with an experimental self-supporting, ferrocement roof dome, which I constructed from a light frame. I then sprayed the roof on before rendering the walls. The weight of the roof precompressed the bales.

Firstly, I poured a combination footing wall and foundation on grade in a circle five metres in diameter on the inside of the wall to a height of 150 millimetres above grade. I embedded plastic garden stakes into this foundation to spike the first course of bales on, then I laid a damp course. I stacked the bales six courses high, leaving room for a doorway and incorporating two windows.

Once the bales were in position I made a top-plate from round pipe with the use of a pipe bender and made a complete ring around the top of the bales. I made a toe up to allow room for a metal door frame and the frame was hung from the top-plate.

I then made a hole under the door frame to allow it to settle with the roof-plate after I sprayed the roof and the walls settled under its weight.

Next I made a light metal frame with a pitch of 32° using a star plate bracket to connect the ten pieces of this frame in the centre of the roof. I connected the outer ends of this frame to the top-plate ring. I then put fencing wire around the roof like a Hill's hoist clothesline. I covered the frame with a layer of chicken-wire, then weed cloth, then chicken-wire again.

Chicken-wire was used to tie the roof and top of the wall together and to tie both the window and door frames to the bales but was not used to cover the whole building.

I sprayed the exterior of the roof with three coats of a cement render mix while the interior of the roof was sprayed with a stabilised earthen render. I then left the building to allow the walls to compress under the weight of the roof.

I sprayed the walls several weeks later, after the walls had settled 200 millimetres. I used an earthen render, stabilised with lime and cement on the exterior walls, while I finished the interior walls with an earthen render with lime as a plasticiser. Fine chaff was added to the render in the second coat to provide some reinforcement throughout the body of the render. I only added about five per cent cement as a stabiliser on the exterior of the building. The entrance way has had full weather exposure for two years with minimal effect on the structure or render.

The Steiner School in Queensland where strawbale wall classrooms were built inside an existing shed.

School buildings

I then went to travel to Queensland to assist in the strawbale construction of four classrooms and a series of storerooms inside an existing shed. When I arrived at the Gold Coast I found that none of the strawbale work had been completed.

However, this didn't pose a problem as it gave me time to do the soil preparation. The job had about 620 square metres of surface area to be rendered, which required about 31 cubic metres of render or 38 tonnes of soil to be prepared, loaded, mixed and applied to the walls. A big job!

The shed had a massive steel frame about five metres high with a full concrete slab floor. The classrooms were to be four egg-shaped rooms running across the shed with open ends to be filled with stud-framed windows and doors.

Each wall was constructed between massive poles wider than a strawbale and stacked on top of a pyrethrum-impregnated damp-course. As the courses were built up wooden garden stakes were spiked down through several courses to tie the walls together. On the final course several centimetres of these stakes were left out to allow the top-plate to be attached.

The top-plate was constructed from green acacia saplings which were harvested from the regrowth forest in the school grounds. These saplings were woven in between the garden stakes and then attached with screws. At any point along the top of the wall there were three to four saplings overlapping, which created a crude, but effective, top-plate.

One run of chicken-wire was placed over the top of these saplings to provide extra reinforcement. Chicken-wire, about 300 millimetres wide, was also used to tie the logs on the ends to the rendered wall, and this was the only chicken-wire used in the building process.

The walls were then rendered with three coats of earthen render, there was no cement used, and the second coat had chaff in it to provide reinforcement. The final coats were done in three different colours — pinky peach, pinky orange and orange yellow. These colours were achieved by adding oxides to a base white clay.

Back to the fence

I finally had time to finish the other half of the fence when I returned from Queensland. This time it proved to be much easier and much cheaper. The footing wall was constructed from recycled cement that had been crushed. During the crushing process about five per cent of the cement is reliberated and when mixed in a cement mixer with a small amount of extra cement it makes a serviceable foundation for a garden wall.

The other cost-saving in time and money was made by placing a single piece of chicken-wire along the top of the wall. I then applied three coats of render in three days, allowing for drying time between coats.

THE KIVA

I was contacted by a neighbour and asked if I would be interested in assisting them in constructing a Kiva, a North American ceremonial chamber. The initial guidelines were that it was to be ten metres (33 feet) long by 6.7 metres (22 feet wide) and 3.3 metres (11 feet) high.

The fact that the Kiva was an oval shape posed many challenges, the biggest was how to build the roof. The eventual solution was a combination yurt and tipi roof made from Corsican pine poles and clad with ecoply sheeting.

Oval foundations were dug by hand, formwork constructed, trench mesh laid and the concrete pour was arranged to construct a one piece foundation and footing wall which extended about 150 to 200 millimetres above grade. Rebar spikes were embedded in the foundation, two per bale, to hold the first course of bales in position.

A door frame was constructed from massive, well-seasoned ironbark sleepers cut from the owner's property many years ago. A double, black plastic damp-course was placed in position, with great care taken to seal it around the rebar spikes and then the strawbale work commenced.

Great care should always be taken when laying bales to ensure that each course is as good as possible, because any errors made on lower courses are often exacerbated as the wall continues. Bales should be placed firmly into position but not forced as this will cause the wall to push out of shape.

Always take the time to shorten a bale when finishing a course: often in a load of bales there will be some short and some extra long bales that can be put aside and used if a full-sized bale will not fit. As the wall was constructed and each course completed each bale was spiked with a wooden garden stake which tied several courses together.

I have found that wooden garden stakes appear to do this better than rebar, they are also cheaper and they take up more space in the bale giving a more stable wall. Stakes with a four-sided point are much easier to drive in than those with a two-sided point which seem to snag on the straw and string as you drive them in.

Rebar may still be the best option for those with termite issues. On the last course the stakes were left sticking out to allow them to be connected to the top-plate. The top-plate was constructed from recycled hardwood which was cut to fit the top of the wall and made a complete ring around the top of the wall to connect the roof poles to. The roof-plate was then tied down to dynabolts in the foundation, using fencing wire and gripples to pretension the strawbales.

At this stage it was decided that a strawbale seat around the inside of the wall would be a handy addition, so a footing wall on grade was constructed for this, a damp-course laid and bales placed in position. Chicken-wire was used to tie the seats and wall together and to tie the doorway to the strawbale wall.

Now the straw was in place with no roof it was very important that the walls were completely covered every night to ensure that they didn't get wet. There is nothing worse than waking up to the sound of rain on your roof knowing that you didn't cover your wall because you thought it wouldn't rain: believe me I've done it.

Having finished the walls, given them time to settle and progressively tightened the gripples it was time to render the walls to give them some stability before the roof was constructed. Three coats of stabilised earthen render were applied to the exterior of the wall and three coats of earthen

The Kiva which, being an oval shape, posed special building challenges.

Right: The studio at Euroa. The owner and Lachlan were pleased with the final result but Lachlan discovered how much more difficult and time consuming it can be to render walls that require scaffolding.

render were applied to the interior. It is important when rendering a wall, especially a loadbearing wall, that both sides of the wall are rendered concurrently. If all three coats of render are applied to one side of a wall it is possible for the weight of the render to pull the wall over.

Once the render had been applied the roof was erected and clad. Due to the pitch of the roof and the roof poles this left a gap around the top of the wall. This gap was covered over with light brown shade cloth to keep insects and birds out and provided under eve ventilation.

STRAWBALE STUDIO

I then moved my equipment over to Mount Tenneriffe, near Euroa, to render a strawbale studio. Like most jobs this had its own unique challenges, the main one being that the straw walls were 3.5 to five metres high. I discovered how much more difficult and time consuming it can be to render high strawbale walls that require scaffolding.

The studio section of this building had been constructed of strawbale on a stone footing wall which looks excellent. The problems with this building started when the strawbales the owner-builder had ordered arrived and were not the quality he had expected.

The problem of poor strawbale quality was then compounded by his much appreciated helpers being a bit over-enthusiastic, rushing ahead and building whole wall sections without ensuring that the courses already built were done properly. Strawbales had also been forced into place which caused the walls to bulge. The result was walls that were quick to erect but it required a lot of time and effort to get the building to the standard the owner and I wanted before rendering.

The other problem caused by loose strawbales is that I needed to use significantly more render to cover the wall, especially with the first coat. The wall had three coats of cement and lime stabilised earthen render, both inside and out. The owner and I were pleased with the final result.

STRAWBALE HOUSE

From here I travelled to Deans Marsh in Victoria's Otway Ranges to help an owner-builder couple render the exterior shell of their strawbale house. I applied three coats to the exterior of the house, with the idea that they would complete the interior by hand. However, once the exterior was done they were so happy with the results — especially the speed of application — that they decided to have the inside done as well.

The render used in this job was an earthen render stabilised with slaked rock lime. The manufacturer claims it is better quality than the readily available hydrated lime and gives greater plasticity and is less prone to shrinkage cracks. The render worked well and seemed less prone to dusting than other earthen renders.

BEFORE YOU COMMENCE

I would like to conclude by making some points which I feel are important to consider before you commence your own strawbale project.

DESIGN AND PLANS

Keep it simple, especially if you or others involved in the project have limited experience with strawbale building. This includes the architect, draftsperson and builders with grand plans. It is important to choose a design that is appropriate for the building material used.

All strawbale walls in Australia should be built with an adequate eaves overhang to protect at least the top of the wall, even when rendered with cement render. The minimum eaves overhang I would consider would be half a metre. My preference would be to have wide verandahs around most, if not the whole, of the building.

The design of the top-plate needs a lot of thought. It is important to design your top-plate to tie your roof, wall, ceiling and eaves lining together to get the finish and look you want. You also need to ensure that the top of the wall has been sealed adequately. My tip would be, if in doubt, make it bigger to ensure it does an adequate job.

Walls above three metres high become significantly more difficult to build, render and finish off, requiring scaffolding to complete these jobs. Give consideration to loft design if you want a second storey.

Keep the number of window and door openings large and to a minimum. Small window and door openings actually add surface areas to the rendering job at the same time making it more difficult and time consuming. Consider full-length windows and doors where practical because then you don't have small straw sections above, and below windows, which can be difficult to both construct and render. However, deep window reveals look good so if that's the look you're after, go for it and do the extra work.

Steel portal frame with strawbale infill is a good combination for strawbale. When you render up to wood — unless it is very hard, seasoned wood — it always absorbs moisture and swells, subsequently shrinking away from the render as it dries. In my experience, this does not happen with steel frames. This also holds true for window and door frames. I am also told that steel and render expand and contract at similar rates which is not the case with wood and render.

When possible window and door frames should be boxed in and the actual windows and doors inserted after rendering. This prevents damage to these items and significantly reduces the work involved in preparation and clean up.

Straw is a super insulator and hence an excellent choice to build the exterior shell of your house. Although strawbales are quite suitable for interior walls they do tend to take up a lot of space in the interior of a house. Their insulation properties are all but lost inside a house and they have minimal thermal mass qualities. For these reasons I am inclined to recommend that interior walls be built from a variety of materials such as stone, mudbrick, rammed earth, straw clay, or even conventional stud frames.

Choosing bales

When sourcing your bales you are looking for a neat, firm consistent bale to provide you with a good quality building block. Over tight bales are: heavy; hard to work with; difficult to shape; difficult to drive stakes through and they also have reduced insulation qualities. Bales that are too loose make for a shabby, poor quality wall.

Ideally bales should be sourced locally from both a cost and environmental perspective. However I don't believe it is worth working with poor quality bales just because they are local. It is much better to spend extra money on a better quality bale which will save you time and money during construction.

Rendering a poor quality wall wastes significantly more time, more money and more materials. Bales should cost between $2.50 and $5 delivered, depending on where you live. Keeping your bales dry is a priority. It is worth getting your supplier to store the bales for you until you are sure you can protect them from the weather.

Most cereal crops — wheat, barley, oats or rice — are suitable for strawbale construction. I prefer wheat because I find them easy to work with. People I know who have used rice have found them to be tough and difficult to work with, rice strawbales also have less insulation qualities as they are solid stemmed.

Rendering

Choosing the render for your strawbale house will be influenced by many things, such as: environmental issues; building design; ease of application; maintenance; aesthetics; availability of materials; the building inspector, and cost.

In my rendering business I have specialised in spraying earthen render with, or without, stabilisation depending on the customer's preference and the surface to be rendered. Many people believe cement is the safe option to render strawbale buildings because it gives greater strength and better water-proofing.

I don't believe this to be the case: it is my opinion that cement is a very brittle, non-forgiving render to work with. In most cases cement requires the additional effort and expense of wrapping the building in chicken-wire to provide reinforcement for this fragile render. Even then you are still likely to get some cracking as your building settles.

Cement is a highly hydroscopic material tending to suck water into your straw wall at the same time reducing the wall's ability to breathe, hence not allowing moisture that may have accumulated in the wall to evaporate out. It is very prone to rising damp problems if your damp-course is inadequate or your design is poor, as it allows moisture to sit at the base of your wall.

Cement is also very high in embodied energy and fairly expensive. I feel it is a shame to build a beautiful alternative house out of strawbales then wrap it in cement. An earthen render is more suited to render a strawbale house and remains in keeping with the theme of strawbales. It is low in embodied energy and most definitely environmentally friendly.

Earthen renders are more flexible than cement and allow the wall to breathe. Earthen renders are easy for a novice to work with and can be repaired relatively easily. There is a degree of dusting with most earthen renders and I would recommend that any earthen render should have a sealer applied. This is best done when the building has time to settle — probably six to 12 months after construction.

There are many variables which need to be considered when deciding on what mix of earthen render to use for a specific job, some of which are as follows:

- prevailing weather conditions;
- amount of eaves overhang on building;
- type of clay to be used;
- mix variation between coats;
- amount of clay in sand used.

A basic rule of thumb for an earthen render is to have 20 to 30 per cent binders to 70 to 80 per cent aggregate. A suitable plasterers' sand for a good quality render is one that has about five per cent clay with aggregate ranging from three to four millimetres down to fines.

First coat (scratch coat)
• 3 parts clay;
• 1 part hydrated plasterers' lime;
• 10 parts plasterers' sand;
• About 40 litres of water.

This makes a rich first coat to provided good adhesion to bales.

Second coat (brown coat)
• 2 parts clay;
• 1 part hydrated plasterers' lime;
• 10 parts plasterers' sand;
• 3.3 litres hammer milled straw (ten to 20 millimetres shattered straw) for fibre reinforcement;
• About 35 litres water.

Third coat (finish coat)
• 2 parts clay;
• 1 part slaked quick-lime;
• 1/3 part off-white cement;
• 1 part fresh cow manure;
• 12 parts plasterers' sand;
• About 30 litres skim milk (casein glue) and/or wallpaper paste as dedusters and sealers.

SEALANTS

For exterior and interior wet areas I would recommend the use of a water-proofing agent once the building has had time to settle, the walls are completely dry and any maintenance has been attended. There are many possible sealers that could be used, both commercial and non-commercial (see *The Straw Bale House*).

WET AREAS

There are several other wet area options which would also work well, listed in my order of preference:

1. Two coats of earthen render, as listed above:
• One to two coats of slaked rock lime render;
• One part rock lime putty to three to four plasterers' sand.

2. Two coats of earthen render, as listed above:
• one to two coats stabilised lime render;
• four parts rock lime putty;
• one part cement;
• 15 to 20 parts plasterers' sand.

3. Two coats of earthen render, as listed above:
• 3 parts hard wall plaster;
• 2 parts rock lime putty;
• 10 to 15 parts of fine sand.

It is important to note that experimentation needs to be done with your own soil before deciding on a recipe.

BENEFITS OF SPRAY RENDERING

During spraying the render is blown into, and compacted onto, the bale, whilst the addition of fibres create a mesh like reinforcement throughout the render. These features significantly increase the strength, density and durability of our renders, virtually negating the need for chicken-wire and other such reinforcements. There is a significant saving on time and labour costs.

A large choice of colour, texture and materials is available in all render finishes including earthen and stabilised earthen renders. Greater continuity of both colour and texture are achieved due to large batching and the ability to finish large areas in one day.

I have access to previously trialled materials and clays, including several excellent white clays which are easily coloured to any colour of the rainbow.

This technique enables me to render in high, difficult and inaccessible areas including window, door and top-plate areas which are very difficult by hand. Rendering time can be reduced from months, and even years at times, to one to two weeks for most houses. (See page 154 for Lachlan's contact details.)

Build your own chookhouse and greenhouse

Thinking of building a strawbale house? Here is a small project to practise on.

by Helen Bernard
Imagine Strawbale Constructions
Daylesford, Victoria.

HENS WILL BE happy and warm in their strawbale home which is warmed by a greenhouse on the north with access to nesting boxes and a room to potter around in. You will need about 90 strawbales, either wheat or rice. The bales should be dry and firmly tied, but they do not need to be specially made for building. Use two-string bales of size: 900 millimetres long by 450 millimetres wide by 350 millimetres. Bring them to the site and stack them in the house once the frame is up, the roof is on, the external mesh is in place, and the walls are wrapped with Tyvek or tarpaulins. This avoids double handling, and reduces the chance of the bales getting wet.

Footings

Make concrete strip footings 450 millimetres wide by 450 millimetres deep. Keep the top of the footings 200 millimetres above ground level and make sure the ground around the house is sloping away from the building. Cover the footings with black plastic, bituminous paint or malthoid to prevent water leaking into bales. Fix two bottom-plates to the footings 450 millimetres apart, and fill the gap between with gravel.

Framing

Posts can be either two 90 by 45 millimetre pine studs nailed together, or 100 by 100 millimetre recycled timber. The bucks are constructed from 90 by 45 millimetre studs set 450 millimetres apart and faced with seven millimetre plywood both sides. A pitching beam, 290 by 45 millimetre F5 pine is bolted to posts and bucks, also serving as a render stop. Rafters are 190 by 35 millimetre F5 pine at 600 mm centres. Roof battens are 35 by 70 millimetres F5 pine at 900 centres. Fix roofing iron over sarking, complete with gutter and downpipe.

Before the straw

Install heavy gauge chicken-wire to the outside of the wall frame, stapling it to the bottom-plate and the top-plate. As well as strengthening the rendered finish, the chicken-wire serves to tie down the roof to the footings. Run the wire from the top to the bottom, overlapping 200 millimetres at joins and ties with O-rings. Fit expanded metal over any timber framing for a good render finish. This may be overdoing it for a chookhouse but it's good practice for the real thing. Wrap the walls with Tyvek and get the bales on site.

Strawbale time

Stack six courses of bales from within the house in stretcher bond (joins between bales do not line up with course above). After laying one course fill gaps between bales with loose straw. Retie bales where shorter lengths are required then cut original strings. Gaps at the top of the walls, or ends where bales meet timber, should be stuffed with loose straw. Once the strawbales are stacked, install chicken

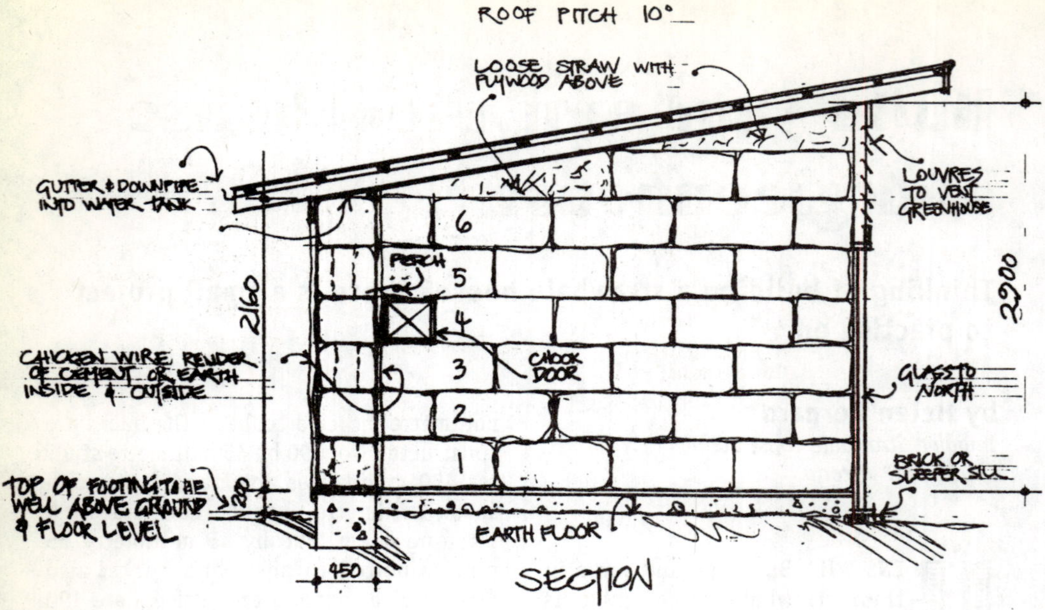
ROOF PITCH 10°
LOOSE STRAW WITH PLYWOOD ABOVE
GUTTER & DOWNPIPE INTO WATER TANK
LOUVRES TO VENT GREENHOUSE
6
PERCH
5
4
3
2
1
2160
2900
CHICKEN WIRE RENDER OF CEMENT OR EARTH INSIDE & OUTSIDE
CHOOK DOOR
GLASS TO NORTH
BRICK OR SLEEPER SILL
TOP OF FOOTING TO BE WELL ABOVE GROUND & FLOOR LEVEL
200
EARTH FLOOR
450
SECTION

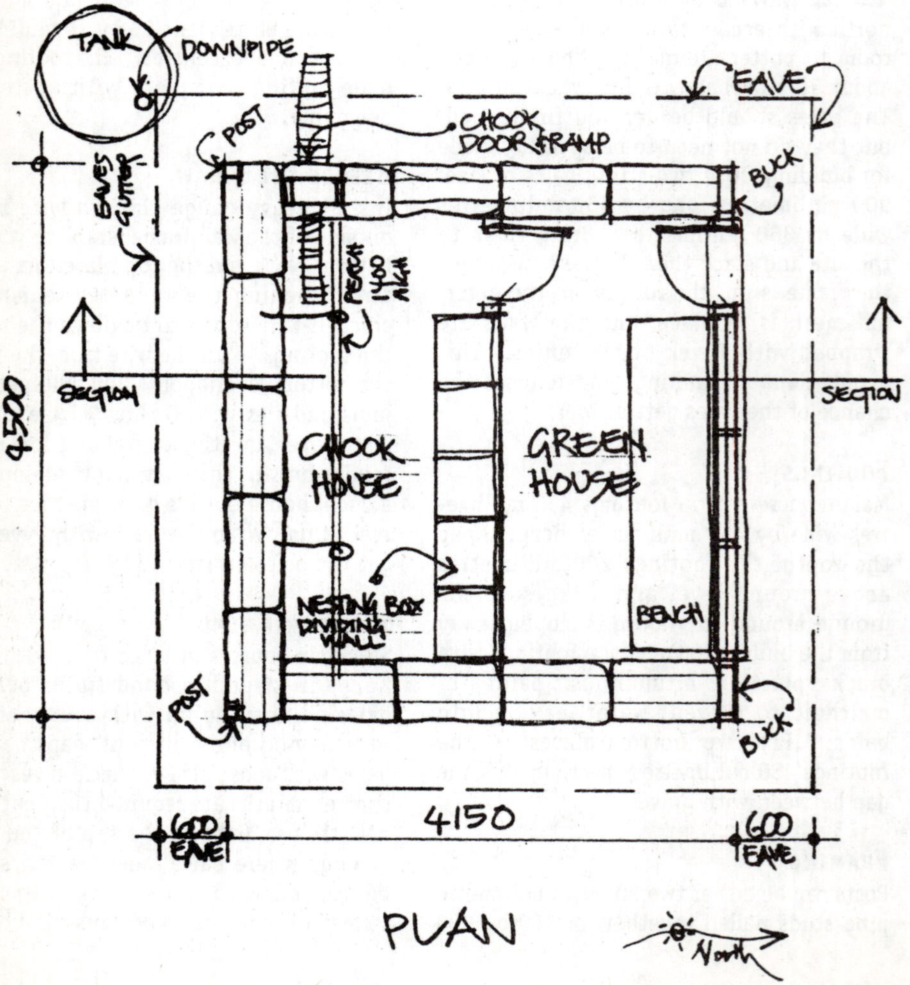
TANK
DOWNPIPE
EAVE
POST
CHOOK DOOR & RAMP
EAVES GUTTER
BUCK
PERCH 1400 HIGH
SECTION
4500
SECTION
CHOOK HOUSE
GREEN HOUSE
NESTING BOX DIVIDING WALL
BENCH
POST
BUCK
4150
600 EAVE
600 EAVE
PLAN
North

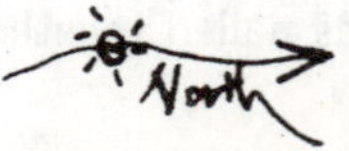

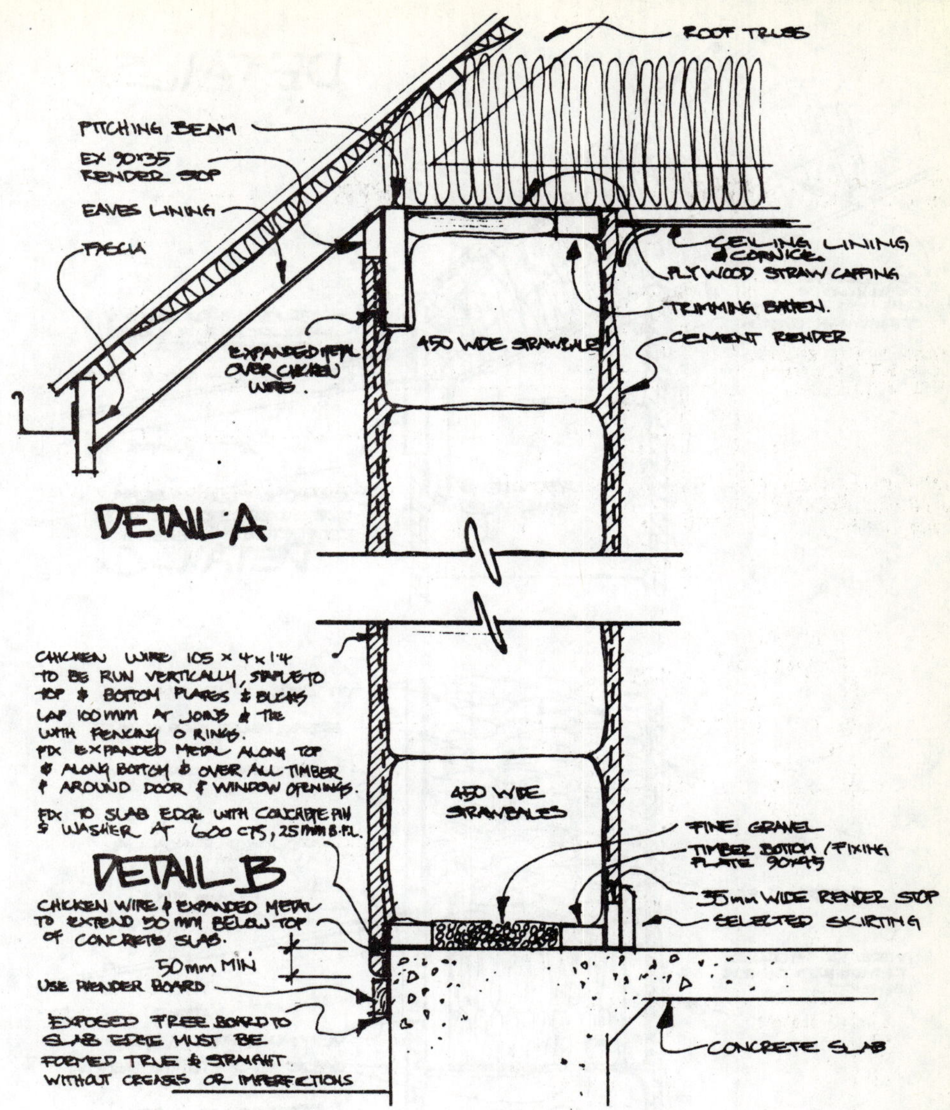

wire on the inside of the walls. There is no need to compress the bales. If some settlement occurs then stuff loose straw above the top course of the bales.

Remove Tyvek from one wall section at a time to stitch internal and external layers of chicken-wire and replace Tyvek on completion of that section. Stitch through the wall with baling twine using a bale needle. This can be made from a 550 millimetre long by six millimetre diameter steel rod with a 150 millimetre angle to form the handle and an eye formed at the other end.

Work in pairs with one person inside and one outside. The stitching pulls the chicken-wire close to the strawbales which reduces the amount of render that is required to cover the wire. The wire can also be tensioned by twisting with pliers, to pull it close to the bales. Once the stitching is complete then fit expanded metal to the inside of the walls.

Any bales which get wet must not be used. If part of a wall gets wet once the bales are in place then the wall must be pulled down and rebuilt using only dry bales.

Rendering

I recommend a cement render for housing because of its durability, however for the

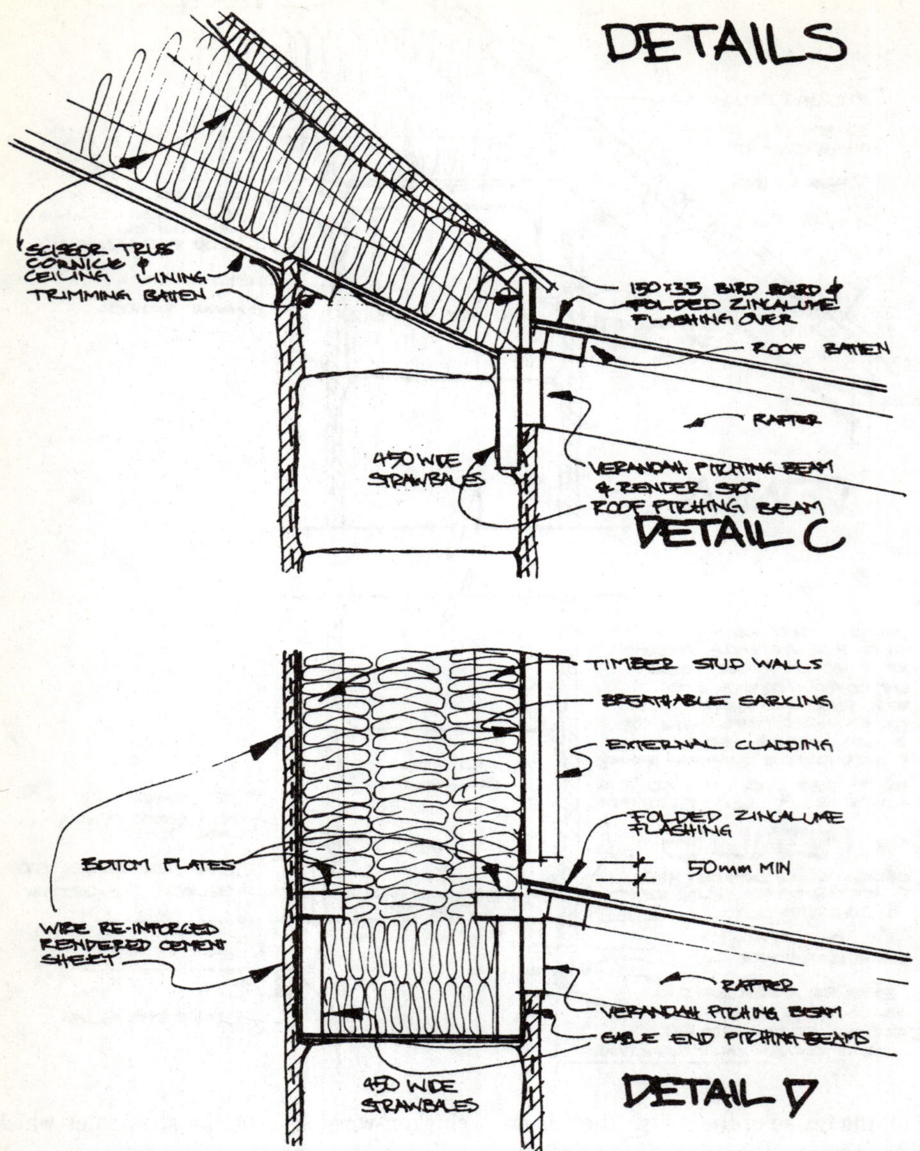

chookhouse you can try clay or lime renders. The finished render thickness is 35 to 40 millimetres, applied in two or three coats. The render can be applied by hand or sprayed on.

Do not use bonding agents. These prevent the strawbale walls from breathing, and will lead to overly high levels of moisture in the strawbales, which can lead to rotting of the bales. There are several ways of getting a coloured finish in a cement render: an iron oxide pigment in the final coat of render; coloured sand (maybe with off-white cement instead of grey) or a finishing coat of limewash.

Perches can be fixed to cement render walls, or to dowels which are embedded in the bales before rendering. A wall of nesting boxes forms the divide between greenhouse and chookhouse. Spread wood shavings over the earth floor for a fine smelling finish.

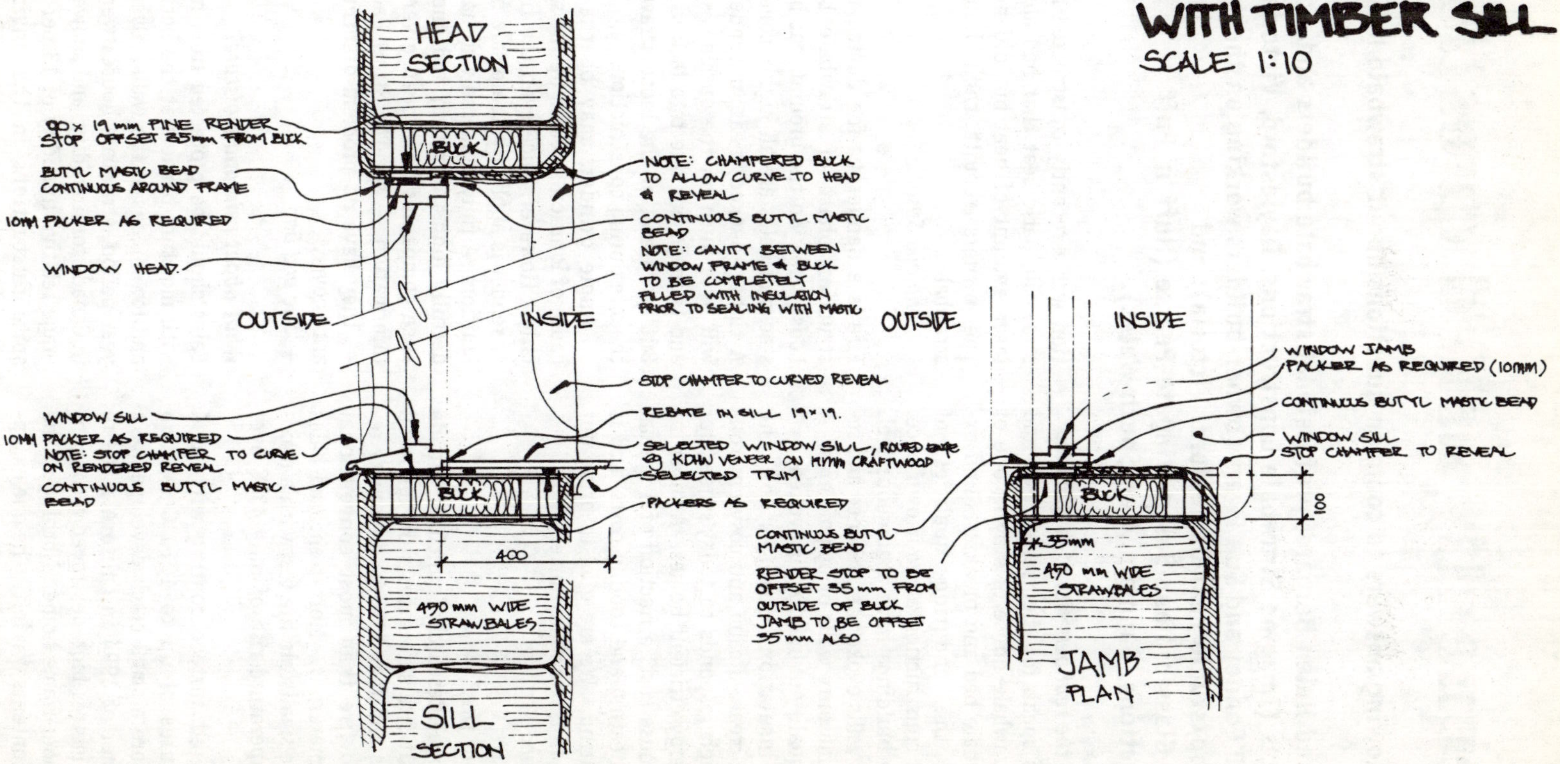
WINDOW DETAILS
WINDOW IN BUCK
WITH TIMBER SILL
SCALE 1:10
HEAD SECTION
BUCK
90 x 19 mm PINE RENDER STOP OFFSET 35mm FROM BUCK
BUTYL MASTIC BEAD CONTINUOUS AROUND FRAME
10MM PACKER AS REQUIRED
WINDOW HEAD
NOTE: CHAMFERED BUCK TO ALLOW CURVE TO HEAD & REVEAL
CONTINUOUS BUTYL MASTIC BEAD
NOTE: CAVITY BETWEEN WINDOW FRAME & BUCK TO BE COMPLETELY FILLED WITH INSULATION PRIOR TO SEALING WITH MASTIC
OUTSIDE
INSIDE
STOP CHAMFER TO CURVED REVEAL
REBATE IN SILL 19 x 19.
WINDOW SILL
10MM PACKER AS REQUIRED
NOTE: STOP CHAMFER TO CURVE ON RENDERED REVEAL
CONTINUOUS BUTYL MASTIC BEAD
SELECTED WINDOW SILL, ROUTED EDGE 9 KDHW VENEER ON MMM CRAFTWOOD
SELECTED TRIM
PACKERS AS REQUIRED
400
450 mm WIDE STRAWBALES
SILL SECTION
OUTSIDE
INSIDE
WINDOW JAMB
PACKER AS REQUIRED (10MM)
CONTINUOUS BUTYL MASTIC BEAD
WINDOW SILL
STOP CHAMFER TO REVEAL
100
CONTINUOUS BUTYL MASTIC BEAD
RENDER STOP TO BE OFFSET 35 mm FROM OUTSIDE OF BUCK
JAMB TO BE OFFSET 35 mm ALSO
35mm
450 mm WIDE STRAWBALES
JAMB PLAN

Frequently asked questions

The following answers to common questions about strawbale building are by:

- **Per and Helen Bernard, professional strawbale builders and architects (Imagine Strawbale Constructions, Daylesford, Victoria);**
- **Don O'Connor and Sue Ewart (owner-builders working on their workshop/studio near Daylesford, Victoria); and**
- **John Glassford and Susan Wingate-Pearse (Huff 'n' Puff Constructions, Ganmain, New South Wales).**

What are the real benefits of strawbale?

The walls speak for themselves. Go and visit a strawbale house or stay in one of the strawbale bed and breakfasts and experience the enduring qualities of strawbale construction at your own pace. Strawbale buildings have a gorgeous feel. The thick walls evoke the presence and security of masonry walls of bygone eras. A timeless quality of building is evoked.

The insulation qualities of the strawbale create a calm and cosy environment which responds to today's need for quiet, energy efficient homes. A super insulated house costs a fraction of a normal home in heating and cooling costs, and is less polluting. We are glad to bring our family up in a strawbale home.

— *Helen and Per*

What about fire?

The National Research Council of Canada carried out fire safety tests on rendered strawbales and found them to be more resistant to fire than most conventional building materials. Mortar-encased bales passed the smallscale fire test with a maximum temperature rise of only 43°C over four hours.

Rendered surface coating withstood temperatures of up to 1010°C for two hours before a small crack developed. In the summer of 1994 in California, a fire storm 90 metres high destroyed a conventional timber-frame home. Nothing of the home or contents was left. The one exception was a recently constructed rendered strawbale patio seat that had absorbed a large amount of heat but did not burn. The owners rebuilt their home with strawbales.

— *John and Susan*

Fire is a dangerous foe during the construction phase of strawbale building. Loose straw on the ground, cut bales and stacks of bales are all highly combustible. A tightly-packed, neat, unrendered wall will only burn on the outside, smoulder and then go out due to a lack of oxygen. But be careful of the loose straw around the site during construction.

Angle grinders may be the biggest cause of ignition so clean up the site regularly. However, once even the first layer of render is applied you are safe from bale ignition. A fully rendered wall has a wonderfully high fire rating and once insurance companies have been given all the information — see The Straw Bale House — they have no problem granting insurance cover.

— *Sue and Don*

What about rain and moisture?

Bales should be protected from moisture, with moisture barriers at the bottom, top and lower sides of the walls. Over a four-year period, strawbale houses were tested in Canada for humidity, and showed an average relative humidity of 13 per cent despite fluctuations in the environment

The remarkable flexibility of strawbale construction: Hilltop Cottage is a two bedroom accommodation house set on stumps on a steep site overlooking Lake Daylesford in Victoria.

around the walls.

The moisture content of straw must be less than 15 per cent. At 20 per cent, bacteria can attack straw, and it may break down. However, most straw that we have measured has been below ten per cent. If this is the case, then rice, wheat, barley and oats are ideal. Rice straw has some advantages over the other cereal straws.
— *John and Susan*

Strawbale walls are best served with generous eaves or verandahs. Keeping the walls as dry as possible will add to the longevity of the building. This needs to be incorporated into the design, while still allowing the sun to enter the building. A carport or garage on the south will protect the southerly wall as well as providing a useful utility area which buffers the building from cold winds. Overhangs to the north should be a compromise between necessary solar gain and light in the building, and some rain protection.
— *Helen and Per*

What about termites?

If allowed access, termites will nest and travel through straw walls to access door jambs, roof trusses and top-plates. For low activity zones, raised dwarf walls are adequate.

For higher activity zones, treatments like Termimesh or Granite Guard or chemicals (please avoid if at all possible) are required as barriers to termite access. In other words, the same precautions as a normal house are required.
— *Sue and Don*

The best protection against termites is to build on a concrete slab. Penetration through the slab for plumbing ducts can be sleeved with a proprietary collar. A freeboard of 150 millimetres of clean and clear slab edge serves the dual purpose of providing good visual detection of any termite galleries which appear, as well as keeping the bales high out of the ground, avoiding moisture problems.
— *Helen and Per*

Strawbales provide fewer spaces and havens for pests than conventional wood framing. If a good coat of render is applied and maintained, access for even small bugs is significantly reduced. Few termites like straw compared with the many that like wood. Clean, bright straw has very few mould or allergy problems.
— *John and Susan*

What about rodents?

Rodents love warm cosy bales of straw. If the render is complete with no holes or cracks to allow them into the interior of the wall, they will not dig a hole through your wall.

There are much easier nests available than having to dig through 40 to 50 millimetres of hard render. However, as we discovered, if you take too long before rendering the bales with at least one coat or you cover the bales with black plastic for long periods of time — beware. This is the way to create an ideal environment for monumental rodent constructions and endless activities.
— *Sue and Don*

How do you get it passed by council?

The State of Victoria, where most of our buildings have been built, has certain freedoms which have made the issuing of building permits for strawbale construction fairly straightforward. Building permits can be issued by private building surveyors, or municipal building surveyors in, or out of, the area in which the building is to be constructed. For Victoria, this has meant that if a local building surveyor is not willing to grant a permit, the permit can be obtained from an alternative source.

Any building technology which is not included in the Building Code of Australia requires certification from a qualified structural engineer. Strawbale construction fits into this category.
— *Helen and Per*

How much does it cost?

Strawbale construction lends itself to community involvement. It is fun to do and encourages group activity. If large families, volunteers and do-it-yourself groups get together and raise some walls, then strawbale construction will become considerably cheaper than any other type of building.

However, if you require skilled tradespeople, then you should expect to pay the same price as, or more than, a standard, brick veneer home.
— *John and Susan*

Some examples of low-cost, strawbale buildings exist. These may be simple buildings, built with voluntary labour, and materials being obtained as cheaply as possible, perhaps even free. They certainly attract a lot of interest. However, if the aim is to build a 20 square family home, with bathroom and ensuite, complying with the Building Code of Australia, it may not be possible to achieve the low prices that sound so wonderful.

The walls of a house may represent 20 per cent of the cost of a building. If strawbale walls were free the saving on the cost of a house, all other things being equal, would be 20 per cent. They are not free. Sure the straw is cheap. Rendering, however, is not. The work is hard, and if you are paying someone else to do it, it is not a cheap wall to build.

If you are budgeting for a strawbale house, talk to someone who has built a similar type of house, with the same level of finish as you require, and built by the same method. If you are going to build the house yourself, then talk to other owner-builders, just as you would talk to former clients of a professional builder you are considering engaging. What is their experience?

Where strawbale building represents savings is in the running cost. Heating and cooling is perhaps 25 per cent of that of a conventionally built house. Over the life of a building, this gives savings of tens of thousands of dollars. Who knows what the cost of heating will be in the years ahead? Environmentally this means

Harry and Mareika Borchard's 18 square house at Stanley in Victoria. "A big needle and a chainsaw are all you need that's special for strawbale building. You don't need to get bogged down in the technical stuff, but paint your bale needle orange so you don't lose it on site!" says Harry.

tonnes fewer Greenhouse emissions are created by a strawbale home.
— *Helen and Per*

Cost is such a variable in strawbale construction. Strawbale buildings constructed by professional builders are generally quoted at under ten per cent, to more than ten per cent, the cost of a conventional building. Savings are then made due to lower heating and cooling costs. Real savings are made by the use of non-skilled labour with construction efforts that don't require muscle. Thus all people (mum, dad, kids, extended family and friends) are able to be involved.

The involvement of many people is one of the joys of this method of construction. A strawbale house is a delight to build, a delight to behold and a very, very fine place in which to live.
— *Sue and Don*

How long will the building last?

There are many fine, solid strawbale buildings still in good condition in the USA from the late 1800s. This can be used as a guide to the possible life of strawbale buildings.
— *Sue and Don*

How do you build with strawbales?

The loadbearing method is one in which the strawbales themselves support the roof load. A hip roof to distribute the roof load evenly onto the walls is ideal for this building method. The amount of carpentry in the wall frame is reduced. It is easier to make gracefully splayed windows openings with this method. You should consider the placement of door and window openings at the design stage, because there is differential settlement between these areas and full height straw wall sections, which will require adjustment of the top-plate. We have designed one loadbearing building, Country Gates cottage. The cottage looks like a traditional Australian country house with verandahs all around, making the building look at home in the environment.

The modified post and beam system uses a timber or steel frame to provide the structure for the building. It is suitable for larger, complex buildings. Building surveyors can be more confident with the modified post and beam system, because the frame can be designed by an engineer using conventional methods. In this system, the bales are merely infill and the main issue for a building surveyor is the durability of the straw.

From a practical point of view, it is very handy to have the roof on before the bales are delivered to the site. This provides a large undercover area to store the bales and to work on them. A tarpaulin can be hung from the frame to protect the bale walls from rain for the few weeks it takes between stacking bales and rendering. With this system rain has not been a problem during the construction.

Posts are made up of pine studs, sheathed in plywood, called bucks. This provides a very strong frame, which braces the wall. These are installed around door and window openings. They support a beam which picks up the roof load or floor joists in two storey construction. This method can be seen in *The Straw Bale House* (page 260) where it is compared to the loadbearing method. Alternatively, a steel frame can be used. The bucks also provide a strong fixing point for curtains and picture rails.
— *Helen and Per*

How do you choose bales?

Straw is the dry, dead ends left after a grain harvest. A suitable range of crops include wheat, rice and barley. Hay bales are not at all suitable because they contain seeds. To date we have used wheat straw as it is readily available from a reliable source. The moisture content should be under 15 per cent.

Generally baling is only done when the straw is dry, otherwise the bales won't keep and will be unsuitable for most purposes, including building strawbale homes. Unless the bales have been stored in the rain, they will most likely be at this level. The ideal building bale is of a consistent length and firmly packed and tied. It is prudent to select the bales yourself.
— *Helen and Per*

What footings are suitable?

Strawbale walls can be built on a concrete slab, strip footings or even on poles, or posts with timber or steel joists. The important thing is that the footing is strong enough to support the weight of the buildings and sturdy enough to avoid cracking the brittle, rendered finish. Concrete footings should be at least 150 millimetres above ground level to keep the bales dry.
— *Helen and Per*

How do you install services?

While this always raises questions, there is in fact no difficulty with installing these services in a strawbale building. There is no extra work for the tradespeople because the walls are strawbale and there should be no extra cost. The usual good building practices for location and detailing of wet areas should apply.
— *Helen and Per*

What can I use for the internal walls?

Any material can be used for internal walls. Ideally they will be thin (thus using up less space), insulative (thus decreasing sound transfer between rooms) and should complement the look of the inside surfaces of the outside walls. Attachment to the outer walls can easily be achieved by driving stakes into them and building these stakes into the interior walls.

Strawbale walls will be thick — great for kids' bedrooms or music studios — but too wide for general rooms. Mudbricks, mudbricks on edge, stud walls or light clay bricks (straw coated in mud and set in brick moulds) walls are all viable options.
— *Sue and Don*

Won't earthen renders wash away?

Earthen renders have been used for a very long time. There are many examples from around the world of earthen renders enjoying a century or more of hanging on.

They are available to everyone. They are ideal for owner-builders everywhere.

Every soil has its own qualities and needs experimentation to find the best recipes and combinations. The use of cement certainly adds stiffness to a rendered wall. If a crack appears in our earthen wall it can be simply repaired with the same earth, re-creating a homogenous wall. Repairing cement renders is not as easy and involves chemical use.

An important part of the building design is to protect the walls with a good roof and eaves but this is not so much of a problem if you live in a low-rainfall area.
— Sue and Don

Cement renders?

Our buildings have all been built in high rainfall areas. We use a cement render, inside and outside. The cement render allows the wall to breathe, so any moisture in the wall can escape. It is important that the finishing colour does not seal the wall, so conventional paints are unsuitable. Colour can be applied with several methods:

- coloured sand;
- an oxide or other pigment in the top coat;
- coloured cement render.

Each of these offers a finish which is maintenance free, without flaking or peeling over time. Alternatively, a limewash paint finish is available in a range of colours. This gives an aged appearance but needs to be re-applied periodically.
— Helen and Per

We do not use cement for render, nor do we recommend it. Earthen renders have a certain warmth and softness. We believe this is an important factor in softening our lives.
— Sue and Don

What's the recipe for render?

People often ask for the recipe for render expecting to find a great secret, one that you have been keeping from the world and will only share with them. There is no such recipe. There is a technique. As each soil, clay and sand has its own unique characteristics the actual amounts that will combine to create the desired results will be unique.

However, the technique that is used to determine the combination of clay and sand for each render layer is one that, once learned, is applied to all situations.
— Sue and Don

Why don't you use chicken-wire?

Chicken-wire is a necessity when using cement renders or when trying to restrain chickens. With cement renders the chicken-wire acts as reinforcement for the render and because you have to attach it to the bales it holds the render in place. Chicken-wire serves no structural purpose with earthen renders.

We do use mesh to shape some areas of the strawbale, particularly where we have a complex shape above a door or window such as an arch or where we have a narrow column of straw next to a support post. A thin strip of mesh is also useful in preventing cracking alongside wood where it meets earthen render. We also use Hessian for this purpose.

When earthen render is applied directly to bales of straw it can be pushed into the bales easily, thus creating a very strong bond. The straws become the reinforcement. From this strongly attached base the other layers of render can then be applied.
— Sue and Don

How long will the render last?

Earthen renders can be rejuvenated by the simple application of another layer. Time between layers depends on the exposure of that particular wall to the elements. It's a great time to change the colour of you house when you re-render the walls.

Just use different clays or add natural colour agents to your clay for a change. Re-rendering can be fun! We currently have a finished wall exposed (no eave) as a sort of test for ourselves and are happy to say it has withstood the high rainfall that our area receives for the last three years.
— Sue and Don

Contact list & further reading

Builders and architects

Name	Details	State, postcode	Contact
Ahtee Chia	architect, Maleny	QLD, 4552	(07) 5499 9115
Michael Kingsbury	designer/engineer, Elamon Cottage, Maleny-Kenilworth Road, Conondale	QLD, 4552	(07) 5494 4817, fax (07) 5494 4817
Frank Thomas	highly-recommended builder, Ebenezer	NSW, 2756	Ph/fax (02) 4579 9573.mob 0408 415 806
Ross Bourne	Gold Coast and hinterland builder	QLD	c/- (07) 3804 0027
Scott McGilchrist	highly-recommended builder, 13 Warialda St, Katoomba	NSW, 2780	<mtnmacs@hermes.net.au> (02) 4782 7583
Ross Young	architect, 79 Loftus Street, Katoomba	NSW, 2780	(02) 4782 2762
McGregor Brothers	Formblock Australia, spray render, earth floors, 133 Stuart St, Katoomba	NSW,2780	(02) 4782 4517
Dean Monahan	Blue Mountains region mudbrick and strawbale builder	NSW	(02) 4567 8045
Bernie Jovaras	architect, Jovaras Westland Partnership, Wodonga	VIC, 3690	(02) 6024 6577
Rick Mitchell	Zone 1 Architecture, Blue Mountains	NSW	(02) 4751 4762
Andrea Wilson	architect, 42 Chisholm St, Greenwich	NSW, 2065	(02) 9439 5140
Simon Hearn	Axis Architecture Services	NSW	(02) 4782 1137
Gary Dorn	Strawbale Constructions, 65 Auckland Street, North Perth	WA, 6006	<garydorn@eepo.com.au> 0419 042 265
Sharron Baker & Rudy Stoffel	Bale Up, Designers and builders, PO Box 637, Toodyay	WA, 6566	(08) 9574 4113
Marcus Ward	architect, PO Box 757, Kyneton	VIC, 3444	(03) 5423 5254
Lachlan Burke	Quick Straw Spray Render, 12 Walker St, Long Gully, Bendigo	VIC, 3550	(03) 54425134, fax (03) 54425156
Peter Scott	architect, 845 Huon Road, Fern Tree	TAS, 7054	(03) 6239 1685
Mark Boyd-Graham	spray renderer, Reign Concrete WaterTanks, Whitlocks Rd, Maldon	VIC, 3463	(03) 5475 1340
Steve Sainsbury	designer and builder, GPO Box 2208 Sydney	NSW, 2001	0417 665 618
John Glassford & Susan Wingate-Pearse	Huff 'n' Puff Construction, Red Cross Hall, Ford Street, Ganmain	NSW, 2702	(02) 6927 6027
Per & Helen Bernard	Imagine Strawbale Constructions, architects and builders, 20 Bridport St, Daylesford	VIC, 3460	<Imagine@netconnect.com.au> (03) 5348 1298
Milosh Obradovic	architect, Hardys Bay	NSW, 2257	(02) 4360 2952
Mark Liebke	builder, Toowoomba	QLD, 4350	(07) 4638 7724
Paul Downton	architect, c/- Ecopolis, Adelaide	SA, 5000	(08) 8224 0981
Nik Vollmer	builder, House of Straw, Lot 16, Albion Heights Drive, Kingston	TAS, 7050	(03) 6229 6900

Open days, workshops, accommodation

Name	Details	State, postcode	Contact
Rose White & John Easterbrook	Otways region strawbale open days and workshops	VIC, 3235	(03) 5236 3218
Graham & Kaye Potter	B&B Country Gates, strawbale Bed & Breakfast cottage, Hindmarsh Tiers Rd, Myponga	SA, 5202	(08) 8554 7222
Bale Up,	Designers and builders, PO Box 637, Toodyay	WA, 6566	(08) 9574 4113
Sue, & Don O'Connor	Gentle Earth Walking, PO Box 395, Daylesford	VIC, 3460	(03) 5348 7506
M & S Hennessy	B&B, Old Leura Dairy, strawbale apartments, 63 Kings Rd, Leura	NSW, 2780	(02) 4782 7239
Maureen Peterson	The Flying Pig, Bed & Breakfast, East St, Daylesford	Vic, 3460	(03) 5348 1222
Huff 'n' Puff Constructions	Red Cross Hall, Ford Street, Ganmain	NSW, 2702	(02) 6927 6027
The Bower	Addison Road Community Centre, Addison Rd, Marrickville, Sydney	NSW, 2204	(02) 9569 7633
Michael Kingsbury	designer/engineer, Elamon Cottage, Maleny-Kenilworth Road, Conondale	QLD, 4552	(07) 5494 4817, fax (07) 5494 4817
Lance Kairl	House of Bales! workshops and construction, PO Box 810, Goolwa	SA, 5214	(08) 8555 4223

Structural Testing

Name	Details	State, postcode	Contact
CSIRO	Building, Construction & Engineering Division, Delhi Road, North Ryde	NSW, 2112	(02) 9888 8888
University of NSW	Building Research Centre, 22-32 King St, Randwick	NSW, 2031	(02) 9398 2233
University of New England	Department of Resource Engineering, Armidale	NSW, 2350	(02) 6773 3776

FURTHER READING

Title	Author	Publisher
The Straw Bale House*	by Athena Swentzell Steen, Bill Steen, & David Bainbridge, with David Eisenberg	published by Chelsea Green
Build It With Bales*	by Matts Myhrman and SO MacDonald	published by Out On Bale
Straw Bale Building	by Chris Magwood and Peter Mack	published by NSP
The Strawbale Solution (30 minute video)	by Catherine Wanek	produced by NetWorks Productions, Inc
The Last Straw (magazine)	Managing Editor: Catherine Wanek, Web site: <www.strawhomes.com>	published by NetWorks Productions, Inc
Buildings of Earth and Straw	by Bruce King	published by Ecological Design Press

Strawbale Internet Discussion List <www.solstice.crest.org/efficiency/strawbale-list-archive/index.html> Hosted by CREST, search by topic

*books available from The Good Life Book Club. Phone (03) 5424 1814 to order, or for free catalogue. <www.goodlifebookclub.com>.

Index

A

Addison Road Community Centre 54
advantages of strawbale 101, 129
Ahtee Chia 26
air pollution from straw burning 4
air texture gun 15
Albany Shire, WA 11
anchoring the roof 33
ant-capping 46, 51, 73
approval process 126, 127
Araluen, NSW 55
Australian Standard 1684 127
aviary mesh 27
Axedale 135, 136

B

Bainbridge, David 3
Baker, Sharron 6, 20
bale needle 7, 106, 121, 145
Bale Up workshop 6, 8
bales
- calculating amount of 120
- care in placing 37
- choosing 141, 151
- compressing 52, 74
- cost 141
- keeping dry 141
- laying 47, 57, 121
- loose 141
- moisture content 120
- over-tight 141
- stitching 121
- straw types 141
- time to buy 120
- trimming 57
- tying down 48
- wheat straw 26

baling twine 121
- to fix wire 30

bamboo rods, for wall pinning 14
base-plate 46
- alternative 56, 57
- for wall 51

beater, for shaping bales 110
beauty of strawbale 3
Beechworth, Vic 108, 112
Bernard, Helen and Per 80, 91, 105
Bill Mollison's cabin 60
bird wire 27
bitumen emulsion,
- as damp course 23

bituminous moisture barrier 51
Blackheath, NSW 72
blue metal infill 12
Blue Mountains, NSW 69
Bondcrete to outside wall 40, 70
bonding agents in walls 146
Brick Tor 7
bucks in wall framing 151
Buddhist lama's residence 61
Building Code of Australia 62, 130
- compliance with 83

building cost 86
building permits 149
building small 117
Burke, Lachlan 90, 134
bushfire 24

C

Caloundra City Council 131
Canowindra, NSW 65
car jacks for compacting wall 78
car tyres for footings 51
carbon credits 61, 62
Carnarvon, WA 6
casein in bonding mix 49
Castlemaine, Vic 132
ceiling insulation,
- recycled pulped paper 106

ceilings, straw 80
cement render 15, 40, 78, 106
- colour 152
- disadvantages 141
- repairing 152

chaff render 74
chainsaw as router 122
chainsaw for notching 14
chicken-wire 7,14,38,52,53,78, 106,111,122,136,137,141,153
- as backing 23
- attaching 30, 58

Childers, Qld 36
chook shed 60
chookhouse 143
choosing bales 141
chopped straw in render 48,100
Clare, SA 82
clay render,
- bonding to lime render 49

clerestory window 26, 108
colour 30,100,111,142,146,153
colour render 106
colouring, ferric oxide 40
compression of wall 99
concrete pump for render 96
concrete raft slab 11
concrete strip footings 23
contract-built home 76
cost 65, 67, 74, 90, 98, 150
costs and savings 97
Country Gates cottage 151
cow manure in render 49, 100
cracking in render 49
cross ventilation 93
Cygnet, Tas 116

D

damp-course 37, 138
- bitumen emulsion 23
- pyrethrum impregnated 137

damproofing walls 120
Daylesford, Vic 91,98,101,105,143
Deans Marsh, Vic 88, 140
Denmark, WA 10
design 44, 102, 114, 140
designing with bales 11
diamond lath 106
diamond mesh, for wall edges 14
door and window frames 122
doors and windows 53, 78, 89
double glazed timber windows 92
double stud frames 12
Downton, Paul 61, 62
drawings for council 127
Duncan, Bill 61

E

earthen render 48,100,134,136-138,140,141
- advantages 141
- brown coat 142
- durability 152
- finish coat 142
- for wet areas 142
- insulation properties 48
- scratch coat 142
- spraying 134

eaves overhang 140, 141
ecoply 138
Ecopolis, Adelaide SA 61
electrical wiring,
- in bale walls 8, 15

Eltham Shire 128
embodied energy 44
energy efficiency 4
Energy Efficiency Victoria 92,107
energy rating 92
engineering certificate 126,127
environmental benefits 4
Environmental Protection Agency 50
Euroa, studio 139
Ewart, Sue 98, 101
existing walls, as formwork 22
expanded metal 106

F

fence, strawbale 134, 135
fire 59
fire resistance 88
fire safety 148
floor, clay pavers 31
flour-paste glue wall sealer 100
footings 46, 50, 83
- chookhouse 143
- rubble 73
- strip 23
- suitable 151
- T-shaped 6

Footscray Secondary College 136
foundations 45
frame 46, 51
framework 36
framing, chookhouse 143
free-flowing spaces 27
fungi in straw render 49

G

Galston, NSW 44
galvanised iron roof 46
Ganmain, NSW 2, 60, 73
Gatton, Qld 33
Glassford, John 2,60,73,80,104
government approvals 11
granite dust, for termites 6
Granite Guard 149
green home guidelines 132
greenhouse 143
gripples 53, 74, 78, 89, 138
gripples, for bale tie down 48

H

Hansons, the 61
health benefits of clay render 49
heating 71, 107, 114, 150
heating and cooling costs 106
hessian 153
history of strawbale 2
Holmgren Design Services 132
homemade render mixer 100
Holmesglen TAFE
- strawbale construction 129

Hornsby Council 44
Howden, Tas 114, 126
Huff 'n' Puff Constructions 50, 61, 73, 104
Hybeam 46
hydrated lime 12, 140
hydronic floor heating 95, 110

I

Imagine Strawbale Constructions 80, 91
infill strawbale 55
infiltration of air 93
insulation 8, 59, 92, 97
insulation, of strawbales 91
insulation properties of earthen render 48

J

jumbo strawbales 61, 103, 104

K

Killara, Sydney 60
Kingborough Council, Tas 126
Kingston, Tas 123
Kiva 138, 139
Kyneton, Vic 132

L

ladder trusses 46
large eaves, reasons for 44
Lethbridge, Vic 103
Leura, NSW 68
Liebke, Mark 60
lime putty 12, 15, 58
lime render 15
- anti-fungal 49

lime-sand render 13
limewash 30
limewash, Porter's 92
Lismore City Council 131
loadbearing construction 3, 94
- design 80
- method 91, 151

local authority, dealing with 126
local government approvals 11
low-impact building 132

M

Maleny, Qld 26
Mary Knapp's chook shed 60
metal frame for walls 33
metal strapping, between posts 37
metal straps to hold walls 38
metal U-pins, for walls 14
methods of construction 3

Index CONT'D

mice 30, 37, 90, 122, 150
mobile crane 26
modified post and
beam construction 11,
91, 92, 94, 97, 151
moisture and rain 148
moisture content 120
Molesworth, Tas 119
Monica Vineyard 103, 104
Mornington Peninsula, Vic 94
mortar placement pump 134
Mount Alexander Shire 105
Mount Tenneriffe 139
mud-based render 48
Mudgee, NSW 61, 66
Mullaloo, WA 17
Myponga, SA 80

N

National Timber
Framing Code 127
niches in walls 30, 48
Nillumbik Environmental
Building Surveyors 128, 129
Nillumbik Shire
Council, Vic 128
non-loadbearing walls 11

O

O'Connor Don 98, 101
office and workshop 124
Orange, NSW 63, 71, 74
orientation 92
Ortech strawboard panels 64
owner-builder, advantages 90

P

Partridge, Harry 60
passive solar design 26, 63
principles of 92
passive solar
strawbale house 132
performance provisions 130
pergolas, for climate control 44
Permaculture Institute,
Tyalgum 60
permaculture movement 3
planning 98, 116
planning permits 150
plans, advice 140
plasterer 16
plasterers' sand 135, 142
plastic compression pipes 21
plumbing 8
allowing for 36
through bales 15
pool fence 60
Porter's limewash 92
post and beam construction 123
post and beam, steel frame 6
precedents, finding 127

Q

quicklime, caution needed 12
Quick-Straw Spray Rendering
90, 134

R

R 3.0 insulation 14
'R' rating of insulation 106
raft slab 11
rafters, roof 14
rain and moisture 148
raising the walls 96
rebales 81
rebar 89, 98, 123
rebar spikes 134, 138
recycled materials 45
render 27, 49
chaff 74
cement lime 8
clay-lime mix 90
coats 15
colour coat 106
lime 7
lime-sand 13
mix 27, 64, 70
mix, external 111
mix, internal 111
spray 80, 107
render mixer, homemade 100
rendering
7, 90, 100, 122, 124, 139
by hand 28
methods 30, 35
over timber 54
Renderlock 64
reobar 6
see also rebar
rising damp 141
Roberts, Bob 61
rodents 149
roof
anchoring 33
framework 14
purlins 52
windows 97
roofing 46
round house 22
rubber sealant, wet areas 21
rubble footings 73

S

sand, plasterers' 135, 142
Santa Fe style 108
sarking 66
school buildings 137
sealants, walls 142
secondhand steel 98
services, installing 151, 152
shed conversion 22
simplest method 22
simplicity of construction 10
site selection 82
slaked lime render 12
slaked rock lime 140
soil cement bricks 7
Solomit straw ceiling 88
South Sydney Council 50
South Sydney Waste Board 50
splash coat of render 48
sponge finish for render 96
spray render 107, 136
benefits of 142
machine 8, 135
Stanley, Vic 110
steel conduit for wiring 46
steel farm sheds 112
steel frame 6
vs timber frame 89
steel garage 36
steel portal frame 98, 140
steel rods to stiffen bales 38
steel spikes 33
stitching the bales 121
Stoffel, Rudy 6, 20
stone footings for walls 84
straw burning, air pollution 4
straw ceilings 80
straw compression 78
straw, types 37
strawbale
advantages of 101,129,148
barley 12
building workshop 12
cost 80
disadvantages of 129
ease of construction 3
history 2, 129
insulation 8
jumbo 61, 104
loadbearing 3, 80, 94
moisture content 120
rice 89
wheat 120
workshop 18, 20
strawbale buildings
durability 151
top ten 60
storeroom 36
strip footings
brick 51
concrete 23
structural engineer's
certificate 11
stud frames, treated pine 12
sustainability 4
sustainable construction 44

T

T-shaped footing 6
Tech Dry 27
Termimesh 12, 149
termite prevention 12
termite tracks 6
termites 149
the Bower, Marrickville, NSW 61
The Last Straw 105
The Straw Bale House 3,10,11,
20,33,36,37,40,55,58,83,94,
97,127,134,142,148,151
thermal mass 44, 92
threaded rod
to hold top-plates 39, 99
tie wire, threading 58
timber framework 78
Timber Framing Code 127
tips for builders 97
tips for success 16
Toodyay, WA 17, 20, 24
Toowoomba, Qld 32, 60
top-plate 137
design 140
from round pipe 136
timber 85
treated pine, poles 83
stud frames 12
trench mesh 6, 23
truth window 80, 92, 109
Tyalgum 131
types of straw 37

U

urban design policies 128

V

vermin 59, 150
see also mice

W

wall trusses 46
walls 78, 120
building 53
compressing 53
cracking 141
damp-proofing 120
horizontal bracing 57
internal 140, 152
preventing sway 57
stabilising 33
vertical pinning 57
Ward, Marcus 132
weather-resistant finish 49
wheat straw 55, 64
cost 55
whipper snipper 14, 48
for trimming walls 7
Whyalla Ecocity Site 61
Whyalla, SA 61
window and door openings 140
window bucks 84
window frames 53
window seats 92
windows 118
recycled 84
windows and doors
full-length 140
windowsills, rendered 58
Wingate-Pearse, Susan
2, 60, 73, 104.
wire staples 78
wire-strainers 78
Wistow, SA 76
wood and render 140
wooden battens
for holding wire 38
wooden dowel
to stiffen wall 39
wooden garden stakes 138
wool ceiling insulation 110
workshops 154
WWOOF 29